Smart Cities

This book seeks to identify and to examine factors and mechanisms underlying the growth and development of smart cities.

It is commonplace to discuss smart cities through the lens of advances in information and communication technology (ICT). The resulting overemphasis on what is technologically possible downplays what is politically, socially, and economically feasible. This book, by analyzing the smart city through a variety of perspectives, offers a more comprehensive insight into and understanding of the complex and open-ended nature of the growth and development of a smart city. A solid conceptual framework is developed and employed throughout the chapters, and a selection of case studies from Europe, Asia, and the Arab Peninsula grants the readers a hands-on perspective of the matters discussed.

The chapters included in this book address a set of questions, such as:

- How do the twin processes of digitalization and smartification unfold in the context of the smart city agenda? How do these processes relate to the concepts of smart city 1.0, 2.0, 3.0, and 4.0?
- In which ways have the spatial aspects of city functioning been influenced by the intrusion of ICT? In which ways do the same processes contribute to the attainment of the UN Sustainable Development Goals (SDGs)?
- What are the implications of smartification and the emergence of smart organizations (public, private, and voluntary) for the spatial development of smart cities?
- Do ICT and its application in the city space boost the processes of revitalization and how does ICT influence the process of gentrification?
- To what extent and how does the intrusion of ICT-enhanced tools and applications in the city space impact a city's relationship with its broader territorially defined context?
- Are the administrative borders and divisions inherent in the fabric of a city becoming less/more porous? How should urban sprawl be conceived in the context of the smart city debate?

This book will have a broad appeal to academics, students, and policy makers with interests in urban planning, sustainable development, cities, economics, technology, sociology, urban studies, digitalization, SDGs, well-being, and resilience.

Anna Visvizi, PhD (dr hab.), economist and political scientist, Professor at the SGH Warsaw School of Economics, and Visiting Professor at the Effat University. Seasoned editor, researcher, recognized author, and accomplished project manager, Professor Visvizi has extensive experience in academia, think-tank and government sectors in Europe and the US, including the OECD. Professor Visvizi's expertise covers issues pertinent to the intersection of politics, economics and information and communication technology (ICT), which translates into her research and publications on applied aspects of ICT, in such domains as smart cities/smart villages, as well as knowledge & innovation management. Professor Visvizi is the Editor-in- Chief of Transforming Government: People, Process and Policy (Emerald Publishing) as well as of Digital Policy, Regulation and Governance (Emerald Publishing).

Hanna Godlewska-Majkowska, PhD (dr hab.), full professor, is Head of Institute of Enterprise, Collegium of Business Administration at SGH Warsaw School of Economics, Poland, and was vice-rector at SGH Warsaw School of Economics in 2016–2020. Professor Godlewska-Majkowska's expertise includes issues in local and regional development, business location, and investment attractiveness of regions, are related to the function of an expert in local government units.

Smart Cities

Lock-in, Path-dependence and Non-linearity
of Digitalization and Smartification

**Edited by Anna Visvizi and
Hanna Godlewska-Majkowska**

Routledge
Taylor & Francis Group

LONDON AND NEW YORK

First published 2025
by Routledge
4 Park Square, Milton Park, Abingdon, Oxon OX14 4RN

and by Routledge
605 Third Avenue, New York, NY 10158

Routledge is an imprint of the Taylor & Francis Group, an informa business

British Library Cataloguing-in-Publication Data
A catalogue record for this book is available from the British Library

ISBN: 978-1-032-53950-8 (hbk)
ISBN: 978-1-032-54249-2 (pbk)
ISBN: 978-1-003-41593-0 (ebk)

DOI: 10.1201/9781003415930

Typeset in Times New Roman
by SPi Technologies India Pvt Ltd (Straive)

Contents

Acknowledgments

Amid increased pace of urbanization, pressures for sustainability, and demands for increased well-being, the otherwise fancy and popular argument on smart cities needs to be rethought. Even if the promise and potential inherent in the application and use of information and communication technology (ICT) in urban space remain as valid as ever, it is necessary to view ICT as a piece and parcel of other processes as a result of which the city, or more broadly, the urban area, develops. This necessitates questions of urban development trajectories and of factors that influence them, as well as questions of how city authorities can meaningfully tap into these processes to either pre-empt risks or capture nascent opportunities. The set of questions that thus emerges is broad and requires a variety of inter-disciplinary approaches, including economic geography, management and entrepreneurship, and politics. In this view, this book represents an invitation to a conversation where – contrary to the mainstream approach – the context in which the smart city develops is examined first to offer leads to a more thorough understanding of processes and mechanisms underpinning the spread, adoption, and utilization of ICT in urban space.

We would like to thank all contributing authors for sharing their research, for their hard work, as well as for their patience during the lengthy editing process. We would also like to thank Professor Higinio Mora (University of Alicante, Spain) for reviewing and endorsing the book. Our gratitude is extended to the following research-supporting entities:

- The Polish National Science Centre (NCN), and grant titled "Smart cities: Modelling, indexing and querying smart city competitiveness" (No. DEC-2020/39/B/HS4/00579), led by Anna Visvizi.
- SGH Warsaw School of Economics, Statutory Research Grant, titled: Smart business – smart regions – smart society: on the way to a new paradigm (1.1. KNOP/S22), led by Hanna Godlewska-Majkowska

Above all, we would like to thank the publisher, Routledge, and the entire Routledge team for the opportunity to collaborate with them. While their professionalism is commendable, we also greatly appreciate their kindness throughout the process.

Anna Visvizi and Hanna Godlewska-Majkowska

Contributors

Shahira Assem Abdel-Razek is currently an associate professor at the Delta University for Science and Technology, Department of Architectural Engineering. She is a member of several research units across her faculty and an active reviewer to several international journals. Her master's and PhD degrees are in the cross-section between architecture, environmental engineering, and public health. Her interests include urban health, user-centered design, sustainable green architecture, and user comfort.

Liliana Andrei is a PhD student, Technical University of Civil Engineering Bucharest (UTCB). Her research interests include autonomous vehicles, mobility planning, and road safety.

Giovanni Baldi is a PhD student, University of Salerno, Department of Scienze Aziendali-Management & Innovation Systems (DISA-MIS). His research interests include smart cities and citizen engagement and customer engagement in tourism and sports.

Antonio Botti, PhD, is Full Professor, University of Salerno, DISA-MIS. His research interests include smart cities, entrepreneurship, tourism management, performance evaluation, and public management.

Emmanuel Juárez Carbajal is a PhD student at Centro de Investigación en Computación, Instituto Politécnico Nacional, México. His research focuses on the development of advanced algorithms for intelligent processing of geospatial information, and geospatial data quality.

Ciro Clemente De Falco, PhD, is Research Fellow, Department of Social Sciences, University of Naples "Federico II". His research interests include algorithms, artificial intelligence, geo-media, and spatial analysis.

Agnieszka Domańska, PhD (dr hab.), is an associate professor at the SGH Warsaw School of Economics, at the Institute of International Studies (ISM), College of Socio-Economics (KES). An economist with a record of research and publications, she is also an experienced lecturer who has taught across Europe and beyond. Her research interests include international economic relations, international business cycle synchronization, transmission of macroeconomic shocks, the effects of international economic policy, the EU transport system, and cooperation between science and business in high-tech industries. She is a permanent member of the following academic societies: INFER; the Polish European Community Studies Association (PECSA); IISES; and the Polish Economic Society. For a number of years, she served as the President of the Staszic Institute. Since 2020, she has been the President of the Polish branch of Principles for Responsible Management Education (PRIME).

Małgorzata Dziembała, PhD (dr hab.), is an associate professor, Head of the Department of International Economic Relations, University of Economics in Katowice, Poland. Her research focuses on European economic integration, European Union regional policy, and competitiveness of the regions, international cooperation of the regions, including cross-border cooperation, and innovation policy.

Joanna Felczak, PhD, is an assistant professor, SGH Warsaw School of Economics, Department of Social Policy. Her research interests include: social policy, education, psychology, and the effectiveness of social services.

Hanna Godlewska-Majkowska, PhD (dr hab.), full professor, is Head of Institute of Enterprise, Collegium of Business Administration at SGH Warsaw School of Economics, Poland, and was vice-rector at SGH Warsaw School of Economics in 2016–2020. Professor Godlewska-Majkowska's expertise includes issues in local and regional development, business location, and investment attractiveness of regions, are related to the function of an expert in local government units.

Sara Mohamed Sabry Zakaria is an assistant professor of architecture at the Faculty of Engineering, Delta University for Science and Technology, Gamasa, Egypt. She is a research fellow at Wessex Institute (WIT), Southampton, UK, and a reviewer at one of their respected journals. She approaches the discipline of transportation planning through spatial studies, sustainable development, and sustainable urban mobility. Dr. Sara is an author of a number of scholarly publications in several topics concerned with sustainable urban mobility, transit-oriented development, or applications of GIS in transportation planning.

Sabina Klimek, PhD (dr), is an assistant professor, SGH Warsaw School of Economics, Collegium of Business Administration, Institute of Markets and Competition. Her research interests include smart cities, women entrepreneurship, SME's entrepreneurship, SME's internationalization, and economic diplomacy.

Paweł Kubicki, PhD (dr hab.), is an associate professor, SGH Warsaw School of Economics, Head of the Department of Social Policy. He specializes in public policy analysis, particularly in disability studies.

Oana Luca, PhD (dr hab.), is a professor and researcher, UTCB, Coordinator of Urban Engineering and Regional Development specialization. Her research interests include: sustainable and smart cities, smart mobility, and smart energy.

Radosław Malik, PhD, Assistant Professor, International Economic Policy Unit, Institute of International Studies, Collegium of Socio-Economics, SGH Warsaw School of Economics, Poland. His scientific interests focus on topics related to offshoring, business service centers, labor market transformations, future studies, and the themes of literature reviews and the use of bibliometric tools in scientific research.

Marco Moreno-Ibarra, PhD, is Titular Professor, Centro de Investigación en Computación, Instituto Politécnico Nacional, Mexico. His research is oriented to geographic information systems, intelligent analysis of geospatial information, and semantic similarity and urban computing.

Abeer S. Y. Mohamed, PhD, FHEA-UK, is a full professor at Architectural Engineering and Interior Design Department, College of Engineering, Majmaah University, Saudi Arabia. She is originally a Full Professor of Building Technology at Tanta University, Egypt, with 26 years of academic and professional experience. She is a leading figure in the field of Building Technology Applications in Architecture & Urban Design with 40 international indexed publications. She was awarded UNEP International – Silver Award for Liveable Communities 2018.

Tomasz Pilewicz, PhD, MBA, is an assistant professor at the Institute of Enterprise, SGH Warsaw School of Economics. His research interests include smart organizations, co-innovation, operations, and teamwork excellence. He is a graduate of SGH Warsaw School of Economics and WU Executive Academy.

Emanuel Răuță, PhD, is Public Policy and Public Services Expert and Senior Associate, Jacobs, Cordova & Associates. His research interests include improvement of regulations and strategic planning for public services provision at national and local level and smart urban development.

Samuel Pérez Rodríguez, MSc, is an associate professor, Department of Biosciences and Engineering, CIIEMAD, IPN, Mexico. His research focuses on pollution of water bodies, water quality, and comprehensive water management.

Emilia Romeo, PhD, is Research Fellow, Department of Business Science – Management & Innovation System, University of Salerno. Her research interests include big data and artificial intelligence, business model, and Lean management.

Magdalena Saldaña-Perez, PhD, is an associate professor, Head of the Laboratory for Intelligent Processing of Geospatial Information Center for Computing Research, Instituto Politécnico Nacional, Mexico. Her research focuses on geospatial data analysis and the use of machine learning for developing smart cities, including social analysis problems.

Ewelina Szczech-Pietkiewicz, PhD (dr hab.), is an associate professor, SGH Warsaw School of Economics, Head of the European Union Unit. Her research interests include smart cities, urban development, and circular economy.

Zofia Szweda-Lewandowska, PhD, research fellow, SGH Warsaw School of Economics. Her main research areas include: population aging, institutional assistance, community care to the elderly, discrimination against older people, active ageing.

Anna Visvizi, PhD (dr hab.), economist and political scientist, Professor at the SGH Warsaw School of Economics, and Visiting Professor at the Effat University. Seasoned editor, researcher, recognized author, and accomplished project manager, Professor Visvizi has extensive experience in academia, think-tank and government sectors in Europe and the US, including the OECD. Professor Visvizi's expertise covers issues pertinent to the intersection of politics, economics and information and communication technology (ICT), which translates into her research and publications on applied aspects of ICT, in such domains as smart cities/smart villages, as well as knowledge & innovation management. Professor Visvizi is the Editor-in-Chief of Transforming Government: People, Process and Policy (Emerald Publishing) as well as of Digital Policy, Regulation and Governance (Emerald Publishing).

Introduction

1 Not only technology

From smart city 1.0 through smart city 4.0 and beyond (an introduction)

Anna Visvizi and Hanna Godlewska-Majkowska

1.1 Introduction: a case for overlapping and complementary perspectives to the study of the smart city

There is an abundance of research on smart cities (Ibrahim & Brahimi, 2024; Nguyen et al., 2022; Appio et al., 2019; Kitchin, 2015). From several perspectives, the literature addresses a variety of issues specific to smart cities' functioning. Nevertheless, as suggested elsewhere (Visvizi et al., 2023; Visvizi & Lytras, 2018), the existing debate on smart cities displays a certain information and communication technology (ICT) bias. That is, advances in ICT, and the intrusion of the latter in the city space, are treated as an uncontestable and a taken-for-granted given, a factor always positively associated with growth and development, and by default with the prospect of citizens' well-being. Of course, several critical – as to the very nature and the implications of the smart city agenda – voices have been raised in the literature, essentially portraying smart cities as a part of a "dreaded" neoliberal agenda (Cardullo & Kitchin, 2019; Grossi & Pianezzi, 2017; Odendaal, 2016). The relevance of these voices notwithstanding, to understand the question of growth, development, and performance in/of the smart city, it is necessary to look beyond ideology. In other words, it is imperative to see the ways in which and to what extent ICT and digitalization, as well as other factors, either contribute to or hamper growth and development in the smart city, thus impacting the prospect of citizens' well-being. This book derives from these assumptions.

Whenever engaging in a conversation on the smart city, it is necessary to make a clear distinction between the smart city viewed as an analytical concept, as a tangible artifact (or a context inhabited by a variety of stakeholders) and as a policymaking objective (Visvizi et al., 2023; Lytras & Visvizi, 2021). The literature abounds with approaches to smart cities, and thus several complementary perspectives exist. In the by now classic take on the smart city (Komninos, 2011), a progression from smart city 1.0 through smart city 4.0 has established itself as a very useful point of reference (LugoSantiago, 2020; Trencher, 2019). In this view, the origin of the smart city and so also the smart city debate are associated primarily with the emergence of ICT, i.e., technology and tools sophisticated enough to be of use by vendors, city authorities, and

DOI: 10.1201/9781003415930-2

eventually also, citizens, in the city space. Indeed, the very possibility of smart city pragmatizing was driven largely by advances in ICT and by the debate on how these could be used in the city. What follows is that the smart city 1.0 approach places emphasis on what is technically possible and what the providers can deliver. In some ways, therefore, smart city 1.0 is a largely technical approach to the smart city, where the interest of vendors is stressed.

As a natural evolution of the concept – in that the vendors certainly need a partner to have respective ICT-based solutions delivered to end-users – the focus of the smart city 2.0 approach leans toward local authorities (Nijkampf et al. 2022; Anttiroiko & Komninos, 2019). Here, the point is to ensure that city authorities are open to and facilitate the adoption of ICT in the city space. This should enable a model of city development, where the adoption and dispersion of technology, or ICT-based solutions, are driven by the city authorities (and not by the vendors). Even in this case though, the nature of the relationship between city authorities and vendors frequently has been called dubious, thus over time paving the way to a number of criticisms toward the smart city "project" (Sadowski & Pasquale, 2015; Calzada & Cobo, 2015; Vanolo, 2014; Gibbs et al., 2013). Against this backdrop, the smart city 3.0 concept has emerged. Consider this, as the notion of accountability is key in modern (democratic) systems, also at the local – including the city – level, the prospect and the possibility of ICT adoption requires citizens' support and involvement. Hence, the smart city 3.0 approach, apart from the intimate vendor–city authorities' relationship, also incorporates the citizen co-creation. Co-creation, defined as the process of diverse stakeholders engaging in practices enabling value creation, or practices necessarily for the city design process (Pellicano et al., 2019), is compliant with the imperative of ensuring equity and social justice (cf. Leclercq & Rijshouwer, 2022). Notably, as the smart city established itself as one of the most salient elements of public discourse and political debates, new arguments and points have been included in the conceptualization of the smart city (Ramírez-Gordillo et al., 2024). Accordingly, the smart city 4.0, the most recent addition to the debate, stresses the self-organizing nature of the smart city. Here, it is argued that the convergence of the real and the virtual contexts creates – under conditions of low cost and high efficiency – opportunities for democracy, participation, inclusion, and empowerment (Visvizi & Lytras, 2019a,b).

That being said, concepts and typologies are useful tools to communicate certain content. Nevertheless, while aggregating certain terms and developments, they also simplify the context, the reality indeed, and, as a result, ignore certain concepts and issues. Therefore, while the smart city 1.0–4.0 approach is useful, it also excludes several categories of meanings, flattens the argument, and leaves several important nuances out of the proverbial bracket. It is perhaps for this very reason that the debate on smart cities thrives, and ever new topics and approaches are being proposed and explored. Clearly, a very important emerging aspect of smart cities research pertains to the disconnect between the technologically feasible, the politically possible, and the usable, i.e., sufficiently meaningful and adequately convenient for diverse groups of citizens to actually consider using it. In the burgeoning body of literature on smart cities, there is an understanding that while technology offers an unprecedented promise in view of city development, much more consideration needs to be given to the questions of (i) how to integrate advances in ICT in city planning; (ii) how to ensure that diverse stakeholders will use it; (iii) how to go beyond the assumption that the inroads of ICT in the city space are linear; and (iv) how to address the still unrecognized challenge of non-linearity of ICT-dispersion in the (smart) city space. The chapters included in this book address these questions. To this end, the argument in this chapter is structured as follows. In the next step, a case for non-linearity and discontinuity in the smart city space is made. Then, the book objectives are discussed. A detailed insight into the book's content follows. Finally, the value added and the key takeaways are presented.

1.2   A case of non-linearity and discontinuity

This book makes a case that non-linearity and discontinuity, embedded in the concepts of lock-in and path-dependence, on the one hand (Choi et al., 2019; Ghitter & Smart, 2009), and in the center–periphery divide (the perspective of regionalism), on the other hand (Horeczki & Pálné Kovács, 2023; Limonad & Costa, 2014), are inherent in processes that foster cities' transition to smart cities worldwide (Gagliardi & Percoco, 2017). Still, the complexity of the issue and, for that matter also the multiscalar conceptual and research challenges it creates, remains underdiscussed in the debate on smart cities. This book highlights this issue and suggests ways of navigating it. Clearly, way more needs to be done if the conceptual and empirical aspects of the issue are to be identified, comprehended, explained, and acted upon.

It is argued that while path-dependence and lock-in define the trajectory of growth and development of several smart cities, non-linearity and discontinuity are equally important, albeit rarely talked about, aspects of smart cities' development. In other words, smart cities are by no means coherent entities in which opportunities and possibilities are evenly distributed, access to services and facilities equitable, and growth and development unfold smoothly. Problems and challenges specific to the classic debate in regional studies (regionalism), including the divide between the center and the periphery, shape the reality of urban areas and cities too (Micek et al., 2022; Di Nucci & Russolillo, 2022). In the context of the debate on smart cities, apart from traditional factors conditioning growth and development, it is necessary to focus specifically on the twin processes of digitalization and smartification.

Admittedly, digitalization, defined as both infrastructure development and the uptake of digital technologies and ICT-based tools by smart cities' stakeholders (citizens, businesses, the public sector), does not spread evenly across urban space and does not progress evenly among the stakeholders. The same applies to smartification, defined as (i) the process of refining a product or a service through integrating in it a digital technology (Schuh et al., 2019) and (ii) the procedure of refining an already existing process (Escolar et al., 2023) oriented toward a more efficient delivery of a given service. In brief, even if the value of ICT is not questioned, several factors weigh in the way, the pace, and the efficiency in which the value and the utility of ICT will be experienced by smart city stakeholders. Thus, while "islands" of ICT-driven growth and opportunities will emerge and will consolidate, several districts/areas, and therefore also several stakeholders, will remain on the less advantaged end of the same process. From a different perspective, smart cities do not emerge in social, political, and/or economic vacuums. While it is tempting to conceive of them as closed entities, in fact they remain open structures, frequently serving as hubs of growth for the geographical areas in which they are located. From this point of view, the discussion on smart cities needs to consider the broader notion of the context, i.e., the spatial/territorial, social, and technological, in which smart cities develop. This realization calls for a broad, inter- and multi-disciplinary approach to the question of smart cities' growth and development.

1.3   The book's objectives, approaches, and analytical perspectives

Considering the thus-defined broad and complex context in which this book is embedded, the book's objective is to offer a critical insight into the variety of factors, issues, and developments that influence the growth and development of smart cities. Considering that ICT is key in any conversation on smart cities, the ICT-driven processes specific to smart city development, including digital-ization and smartification, are factored in the equation. However, an explicit attempt is made throughout the chapters included in this book to reflect on the role economic and spatial factors play in the complex process of a city's transi-tioning to acquiring features of a smart city. In other words, the chapters included in this nook address the following objectives:

- Which developments in the city/urban space today attest to the mechanism of path-dependence and lock-in as regards growth and development, and what role may ICT have played in this context?
- How do the twin processes of digitalization and smartification unfold in the context of the smart city agenda? How do these processes relate to the con-cepts of smart city 1.0, 2.0, 3.0, and 4.0?
- In which ways have the spatial aspects of city functioning been influenced by the intrusion of ICT into the urban space? In which ways do the same processes contribute to the attainment of the UN Sustainable Development Goals (SDGs)? (Visvizi & Perez del Hoyo, 2021)
- What are the implications of smartification and the emergence of smart organizations (public, private, and voluntary) for the spatial development of smart cities?
- How should spatial management be conceived in smart cities? How should the spatial limitations of smart cities' growth be dealt with? What role may ICT play in this respect? In which ways and to what extent does ICT influence the processes of linearity and non-linearity in a city's growth and development?

- Do ICT and its application in the city space boost the processes of revitalization (economic, architectural, infrastructure-related, and other) and to what extent? How does ICT influence the process of gentrification?
- To what extent and how does the intrusion of ICT-enhanced tools and applications in the city space impact on a city's relationship with its broader territorially defined context?
- Are the administrative borders and divisions inherent in the fabric of a city becoming less/more porous? How should urban sprawl be conceived in the context of the smart city debate?

1.4 The book's structure and the issues addressed

Apart from this introductory chapter, the remaining 12 chapters have been divided into three parts. Part I (Chapters 2–5), titled "Spatial aspects of smart cities' growth and development," delves into the conceptual issues pertaining to the discussion in this book. Part II (Chapters 6– 8), titled "Territory, scale, inclusion, and participation in the smart city debate," engages with the broader context in which smart cities are embedded. The third part (Chapters 9–13), titled "Navigating the constraints of time, space, territory, and built environment in the smart city context," suggests ways in which selected aspects of non-linearity in the smart city growth and development can be mitigated.

Part I of the book opens with Chapter 2, written by Hanna Godlewska-Majkowska. By applying the economic geography perspective, this chapter examines the mechanisms that shape the spatial structure of (smart) cities, thus influencing the prospect of growth and development of a city (viewed as an aggregate concept) and of specific areas within a city. It is argued that up to 50% of the cost of manufacturing goods and services depends on the characteristics of the location. The prospect of locating the process of manufacturing goods and services in a given city/urban area is associated with increased growth and development opportunities, i.e., through investment, employment, etc. And yet, the intrusion of ICT in urban space, and, hence, smartification of cities, makes it necessary to rethink and redefine the set of characteristics of a given, considered beneficial, location. Given that ICT may serve as a lever of enhanced growth and development opportunities, the important question is how ICT and ICT-based tools and applications specific to the concept of the smart city weigh in the equation of a location attractiveness. An equally important question is that of how to implement ICT-based tools, or how to promote smartification of the city space, in a manner that will add to, and possibly create synergies with the spatial development strategies. The possibility of addressing these two questions is contingent on navigating a conceptual challenge. That is, a relative mismatch exists between the precepts of traditional spatial development models, on the one hand, and the on-the-ground reality of cities adopting ICT-based tools and applications, on the other. The, embedded in economic geography, traditional approaches to spatial development, focus (in both prescriptive and descriptive manner) on the question of the (most efficient) distribution of resources (and the mechanisms underlying it) in the city. Conversely, the mainstream debate on smart cities focuses on ways and modes of harnessing advances in ICT to boost, what clearly amounts to over-simplification, quality of life in the smart city. Since both debates unfold at, essentially, different levels of analysis and display different ontological positions, only a careful bridging of these two approaches is possible. To this end, Chapter 2 demonstrates how the path-dependency theory and the concept of lock-in may be helpful bringing the conversations on spatial development and smart cities closer together.

Chapter 3, written by Malgorzata Dziembała, Radosław Malik, and Anna Visvizi, explores the smart city as a part of a broader geographical, or regional, context, in which it is located. As the authors argue, the smart city is not located in a vacuum. As a provider of services, it provides the local community, i.e., city inhabitants. However, smart services, provided mostly online and defined elsewhere as "pre-emptive," thus seeking to address a complementary need that may emerge while delivering the key service sought (Malik et al., 2021), are already available to anyone interested in them, regardless of their location. This posits a very interesting question of how, on the one hand, the intrusion of ICT in the fabric of the city and, on the other hand, the processes of digital-ization and smartification influence the already consolidated territorial divides, i.e., those between the urban and the rural, and between that which is in the administrative boundaries of the city and that which is not. An argument is made that an ICT-supported emergence of a virtual space, where a great num-ber of smart services can be delivered, at some point will change our under-standing of what cities are and what they are for. To address this complex and multifaceted issue, the discussion in this chapter is framed by a set of overlap-ping concepts, including smart territory, smart services, smart services' ecosys-tem, cohesion, regional disparities, digitalization as well as other.

In Chapter 4, written by Abeer S. Y. Mohamed, the Line City (Saudi Arabia) is examined as an example of a smart city, where the challenges of smartifica-tion, quality of life, and urbanism overlap need to be addressed. As the author argues, the smart city paradigm is the source of a major shift in urban studies, as well as a transformation engine of urban technological innovation. The lat-ter, increasingly, is driven by environmental considerations. The rapid and con-tinuous development and intrusion of ICT in city life have pushed the city on a smartification path. In this context, the imperative of quality of life has emerged as the central tenet of urban development. The onset of the quality of life imperative, most closely associated with the need to ensure person–environment interaction in an urban context, gradually leads toward the emergence of a new concept and a new logic in urban design, i.e., the empathic cities logic. Against the backdrop of the case of the Line City, this chapter conceptualizes the pro-cess of attaining the quality of life imperative through urban smartification.

Against the backdrop of the case of Alexandria (Egypt), in Chapter 5, Shahira Assem Abdel-Razek and Sara Mohamed Sabry Zakaria explore the notion of discontinuity in city development. As the authors argue, the continuous evolution of the urban space remains a niche topic, mostly because research-wise, it is very challenging to capture the co-existing processes of growth and rapture in a city's evolution. The rapid pace of urbanization and, correspondingly, of urban decay bear implications for access to services and facilities. The objective of this chapter is to contribute to the existing debate on means to provide a better and more balanced urban environment where access to nature and recreational spaces is not a privilege but a fundamental right for all residents, regardless of their socioeconomic status. More importantly, it offers a new perspective on addressing the equity crisis in public green open spaces, capitalizing on the urban fabric's inherent rifts. By reimagining urban discontinuity as a resource rather than a hindrance, this chapter opens avenues for inclusive urban planning striving for sustainable city development.

Part II of the book, opens with Chapter 6, written by Tomasz Pilewicz, the question of smartification of public space management and its implications for cities' growth and development are examined. Specifically, by focusing on space management and infrastructure investment planning practices in selected cities in the European Union (EU), this chapter delves into the intricacies of de facto smartification of these cities. Against the backdrop of literature review, analysis of official Internet portals, an online survey (n=25), and individual direct interviews with city representatives (n=9), selected practices related to smartification of public space management, are identified and discussed. In brief, this chapter offers insights into the variety of solutions and practices that, while employed by city authorities across the EU, advance public space management. As such, they are however a source of both benefits and risks. The findings suggest that the benefits include increased citizens' engagement, socio-economic impact, co-responsibility for local development processes, and lower costs of information for citizens and enterprises.

Chapter 7, written by Ciro Clemente De Falco and Emilia Romeo, the question of safety – viewed through the lens of algorithms and geo-discrimination risks – is examined. In the context of smart cities, people and places get connected by means of sophisticated technologies, tools, and applications, such as for instance data mining, machine learning, big data, and the Internet of Things (IoT). The development of various models like forecast, preparation, and monitoring in smart cities has been enhanced by deep-learning and machine-learning techniques for urban development. In self-driving cars, traffic lights and traffic control systems, systems to support the management of energy distribution, and water – in sum, wherever there will be large masses of information to be managed – there is an opportunity to develop algorithms to handle it. Although presented as objective tools, algorithms are to be regarded as sociotechnical devices that incorporate the perspectives of the authors involved in their creation. Thus, mentioning "algorithmic risk" is important to highlight the potentially harmful consequences – intentional or unintentional – that algorithms can generate for individuals and societies in smart cities. The objective of this chapter is to examine a possible hypothetical "geo-discrimination risk," an issue that might transform smart cities in places where opportunities and possibilities are not evenly distributed. The findings of this research suggest that the unquestioning integration of algorithms in the urban development of smart cities has the potential to affect both the physical urban environment and the lives of individuals, leading to the emergence of geo-discrimination at both of these distinct levels. Accordingly, this work may inform future research in this area and support policymakers in developing and managing smart cities.

Chapter 8, written by Marco Moreno-Ibarra, Magdalena Saldaña-Perez, Samuel Pérez Rodríguez, Emmanuel Juárez Carbajal, discusses the impact of generative artificial intelligence (GenAI) on smart cities, especially on their efficiency, cohesion, and sustainability. Given the variability of challenges cities and urban areas face today, including most profoundly inclusion, safety, and resilience, the objective of this chapter is to examine how GenAI may facilitate the process of addressing these challenges, thus fostering sustainable cities. By focusing explicitly on smart cities, in this chapter, it is argued that a great variety of GenAI-based applications exist, and these may bear great value for smart cities and their inhabitants. Indeed, by now GenAI and GenAI-based applications have proven useful in integrating and analyzing information, thereby facilitating operations such as analysis, data management, and description. Accordingly, when utilized in the smart city and/or urban context, GenAI is of value in relation to urban planning, disaster preparedness and mitigation, traffic management, public safety, waste management, and, finally, improving the efficiency and quality of services' provision to the city inhabitants. Clearly, the use of GenAI in this context presents challenges too, including data timeliness, different representations, and other issues related to the complexity of urban data. These considerations should be considered in the design, development, and implementation of GenAI-based applications to ensure their effectiveness, accuracy, and compliance with regulatory frameworks as well as with values and principles. None of these is possible without an explicit engagement of human beings, i.e., experts, who would provide monitoring and oversight over the GenAI-devised suggestions and recommendations.

Part III of the book starts with Chapter 9, written by Ewelina Szczech-Pietkiewicz, Zofia Szweda-Lewandowska, Joanna Felczak, Paweł Kubicki, examines the smart city through the lens of elderly people and the opportunities to access, use, and benefit from the smart amenities. The question of inclusion (or exclusion) and dignified housing conditions play an important role in the argument developed in this chapter. As the authors argue, public space and living conditions determine the possible activity of an elderly person outside the household. They also influence the possibility of living in the given milieu and impact the quality and comfort of life. The use of ICT-based tools and solutions, both in the context of built environment (BE) and at home, may have substantially influenced their safety, sense of security, and overall quality of life. These conditions increase their impact on urban design and management with the growing challenge of an aging society. Against this background, this chapter examines how ICT may be useful in the process of addressing the twin challenges of aging population and evolving, notably becoming ever smarter, i.e., infused with ICT-based solutions, urban environment. The methods used in this study include literature review, mostly used to set research questions and hypothesis, and in-depth interviews. The latter method allowed authors to verify the role of identified factors in creating an age-friendly smart city in three areas, i.e.: policy making, future vision of cities, and necessary actions. The analysis brought results that can be grouped as policy instruments, best practices; and success and failure factors in building urban environments that are supporting and inclusive to all demographics. Implications allowed to create a series of recommendations for the use of smartification and digitalization in responding to the challenge of ageing.

Chapter 10, written by Agnieszka Domańska and Radosław Malik, explores the role of smart transport systems in the process of facilitating growth and development in smart cities. This chapter delves into the discourse on spatial order and the evolution of smart cities, offering insights into three research themes that pertain to non-technology-centric facets of smart city expansion. Initially, a semi-systematic literature review on the implementation of the smart city concept is employed as a research methodology to delineate the current advancements in the development of smart cities in Poland. This review reveals that numerous commendable outcomes have been realized in this domain, predominantly attributed to initiatives backed and jointly funded by the European Union (EU) Operational Programs in Poland. Nevertheless, these accomplishments do not address the central issue faced by Polish cities, i.e. a poorly managed urban development strategy resulting in excessive urban density and escalating traffic congestion. Subsequently, intelligent transport systems are explored as an emergent factor influencing the spatial growth of smart cities and a prospective solution to urban sprawl and other impediments to sustainable urban expansion. In conclusion, the introduction of Intelligent Transport Systems (ITSs) in Poland and their consequent outcomes are showcased as potential catalysts for urban revitalization and enhancers of the spatial dimensions of economic cooperation. This chapter posits that for a comprehensive approach to the spatial growth of smart cities, an amalgamation of non-technology and technology-driven solutions might yield nearly optimal results.

In Chapter 11, written by Liliana Andrei, Oana Luca, and Emanuel Răuță, the role of automated vehicles in smart cities, including challenges and impact on growth and development, is examined. Autonomous vehicle research in Romania is in its very early stages. A limited number of scientific papers have been published on the topic, investigating different factors influencing mobility behavior related to autonomous and connected transport (ACT) systems and in relation to the spatial development patterns of large cities in Romania. There is currently an abundance of smart city strategies in Romanian cities, which are not correlated or aligned with sustainable urban mobility plans or general urban plans, and none of them discuss the presence of ACT. The national ITS strategy does envisage the introduction of ACT, yet it is not indicating how its elements will be integrated into SUMPs. Considering all the above, this explores the relationship between autonomous vehicles and urban structures. A set of 20 interviews was conducted with experts in the fields of ITS, urban and transport planning and local government, on the urban policy framework, the relationship between automated vehicles and urban structure, the determination of barriers to the implementation of ACT in urban areas, and measures to overcome them. Results showed that although there is a potential risk of urban sprawl caused by ACT convenience, automated vehicles have a high potential for deployment in Romanian cities for public transport and other forms of shared transport, also on last mile logistics, showing similarities with the findings in international literature and advancing the concept of the smart city in Romania. Obstacles can be overcome through a better understanding of disruptive factors and constructive collaboration between all stakeholders.

Chapter 12, written by Sabina Klimek, examines the question of securing funding of large, costly, with a delayed return on investment ICT infrastructure development projects. By focusing on the case of Istanbul, this chapter examines whether and how the introduction of ICT and ICT-based solutions in Istanbul influences gentrification, the shape, development, and growth of the city and its population, as well as the attractiveness of the city as a business, scientific, and tourist destination. At the same time, based on an analysis of recent investments schemes implemented under the public–private partnership (PPP) models in Istanbul, this chapter examines their impact on Istanbul's development as a smart city. It is argued that PPP, viewed as a tool for implementing modern solutions in cities, is a highly efficient tool, as it enables city authorities to implement complex, expensive, and technologically extensive changes in the fabric of the city in a relatively short period of time. Substantial positive outcomes for the quality of life of the city residents are thus created.

In Chapter 13, Giovanni Baldi and Antonio Botti delve into the question if smart cities need to be also fast paced and relatively big. Indeed, small towns frequently feature a slow pace as regard how things are done and how life goes by. The latter tends to be interpreted as a slow and healthy lifestyle. However, there is a catch. Given the pace of processes of smartification and digitalization taking place elsewhere, the lax pace of developments, along with weak infrastructure, cultural issues, and a lack of skills, may turn into a significant barrier to growth and development. The question is, therefore, given the specificity of small towns, which factors and mechanisms are conducive to promotion of smart (or, perhaps wise?) sustainable small town/city development. The findings of the research suggest that similarly as in the case of typical smart cities, also in small smart towns, smart public infrastructure, smart mobility, smart energy efficiency, safety are considered crucial in view of attaining prosperity and well-being, regardless if the pace of living is fast or slow.

1.5   Conclusions and takeaways

This book is the outcome of the realization that the debate on smart cities may be booming. Yet, depending on the research angle adopted, or perhaps even more precisely, depending on the disciplinary "home base" of respective participants of the debate, smart city is viewed differently, discussed differently (also in terms of concepts), and a set of different challenges is identified. Indeed, the key concepts defining the discussion in this book, i.e. non-linearity and discontinuity, path-dependence and lock-in, rupture, and change, etc., raise different associations across disciplines. Recognizing this challenge, the authors contributing to this book were encouraged to focus on representations of these concepts, as relevant in their respective research domains, e.g., be it urban design and BE (Chapters 4 and 5), be it access to public domain services (Chapters 6, 7, and 8), economic growth and business opportunities (Chapters 2, 3, and 12), mobility and connectivity (Chapters 5, 10, and 11), and accessibility, inclusion, and well-being (Chapters 9, 12, and 13). By so doing, i.e., by showcasing real developments and real challenges, e.g., as experienced by elderly people (Chapter 9) or by the city authorities seeking access to finance (Chapter 12) and so on, the chapters included in this book reveal the process of navigating non-linearity and discontinuity as they remain ingrained in the city space regardless of and because of the intrusion of ICT and ICT-based solutions in the city fabric. In this way, this book offers a comprehensive and a very sober view of the smart city today. In this book, the smart city is seen as a desirable development, nevertheless a process beset by a variety of challenges. Opportunities exist, yet these need to be handled very carefully, drawing from insights from a variety of stakeholders, possibly capitalizing on the mechanism of co-creation.

The book takes a broad and comprehensive, yet conceptually disciplined, approach to smart cities and challenges to their growth and development. Apart from a conceptual backing, the book features several case studies, including the city of Alexandria, the Line City, Istanbul, as well as several European cities. The focus is directed at diverse stakeholders, i.e. elderly people, the business sector, the providers (vendors), city authorities, as well as decision-makers. Recent advances in ICT, including ChatGPT and the metaverse, are also included in the book. In this way, the book offers a sharp focus on challenges to smart cities' growth and development, a topic that while still nascent will gradually occupy an ever-more-important place in the debate on smart cities and digital transformation. This book offers an understanding of problems and challenges associated with the development and operation of smart cities. In this way, it also organizes the knowledge of smart cities and may add to an increase in the overall awareness of what smart cities are and what the debate on smart cities entails. As such, the book is aimed at a variety of readers, including policymakers, managers, public sector employees, smart cities experts, economists, academics, historians, and students. This book is dedicated to all those interested in smart cities, citizen engagement, sustainable development, growth, development, and well-being.

References

Anttiroiko, A.V., Komninos, N. (2019) Smart Public Services: Using Smart City and Service Ontologies in Integrative Service Design. In: Rodriguez Bolivar, M.P. (eds) *Setting Foundations for the Creation of Public Value in Smart Cities. Public Administration and Information Technology*, vol 35. Cham: Springer. https://doi.org/10.1007/978-3-319-98953-2_2

Appio, F.P., Lima, M., Paroutis, S. (2019) Understanding smart cities: Innovation ecosystems, technological advancements, and societal challenges, *Technological Forecasting and Social Change*, 142, 1–14. https://doi.org/10.1016/j.techfore.2018.12.018

Calzada, I., & Cobo, C. (2015) Unplugging: Deconstructing the smart city. *Journal of Urban Technology*, 22(1), 23–43. https://doi.org/10.1080/10630732.2014.971535

Cardullo, P., Kitchin, R. (2019) Smart urbanism and smart citizenship: The neoliberal logic of 'citizen-focused' smart cities in Europe. *Environment and Planning C: Politics and Space*, 37(5), 813–830. https://doi.org/10.1177/0263774X18806508

Choi, C.G., Lee, S., Kim, H., Seong, E.Y. (2019) Critical junctures and path dependence in urban planning and housing policy: A review of greenbelts and New Towns in Korea's Seoul metropolitan area. *Land Use Policy*, 80, 195–204. https://doi.org/10.1016/j.landusepol.2018.09.027

Di Nucci, M.R., Russolillo, D. (2022) Energy Governance in Italy. In: Knodt, M., Kemmerzell, J. (eds) *Handbook of energy governance in Europe*. Cham: Springer. https://doi.org/10.1007/978-3-030-43250-8_16

Escolar, S., Rincón, F., Barba, J., Caba, J., de la Torre, J.A., López, J.C., Bravo, C. (2023) A methodological approach for the smartification of a University Campus: The smart ESI use case. *Buildings*, 13(10), 2568. https://doi.org/10.3390/buildings13102568

Gagliardi, L., Percoco, M. (2017) The impact of European Cohesion Policy in urban and rural regions. *Regional Studies*, 51(6), 857–868. https://doi.org/10.1080/00343404.2016.1179384

Ghitter, G., Smart, A. (2009) Mad cows, regional governance, and urban sprawl: Path dependence and unintended consequences in the Calgary region. *Urban Affairs Review*, 44(5), 617–644. https://doi.org/10.1177/1078087408325257

Gibbs, D., Krueger, R., MacLeod, G. (2013) Grappling with smart city politics in an era of market triumphalism. *Urban Studies*, 50(11), 2151–2157. https://doi.org/10.1177/0042098013491165

Grossi, G., Pianezzi, D. (2017) Smart cities: Utopia or neoliberal ideology? *Cities*, 69, 79–85. https://doi.org/10.1016/j.cities.2017.07.012

Horeczki, R., Pálné Kovács, I. (2023) Chapter 13: Governance challenges of resilient local development in peripheral regions. In *Resilience and regional development*. Cheltenham, UK: Edward Elgar Publishing, https://doi.org/10.4337/9781035314058.00021

Ibrahim, A., Brahimi, T. (2024) Mapping the research landscape of social and cultural impacts on smart cities. In: Visvizi, A., Troisi, O., Corvello, V. (eds) Research and Innovation Forum 2023. *RIIFORUM 2023. Springer proceedings in complexity*. Cham: Springer. https://doi.org/10.1007/978-3-031-44721-1_10

Kitchin, R. (2015) Making sense of smart cities: addressing present shortcomings. *Cambridge Journal of Regions, Economy and Society*, 8(1), March 2015, 131–136. https://doi.org/10.1093/cjres/rsu027

Komninos, N. (2011) Intelligent cities: Variable geometries of spatial intelligence, *Intelligent Buildings International*, 3(3), 172–188. https://doi.org/10.1080/17508975.2011.579339

Leclercq, E.M., Rijshouwer, E.A. (2022) Enabling citizens' right to the smart city through the co-creation of digital platforms. *Urban Transform*, 4, 2. https://doi.org/10.1186/s42854-022-00030-y

Limonad, E., Costa, H. (2014) Edgeless and eccentric cities or new peripheries? *Bulletin of geography*. Socio-economic series. Online. 13 June 2014. No. 24, pp. 117–134. [Accessed 7 February 2024]. https://doi.org/10.1515/bog-2014-0018

LugoSantiago, J.A. (2020) Is there such a thing as the smart city 1.0, 2.0, or 3.0? In *Leadership and strategic foresight in smart cities*. Cham: Palgrave Macmillan. https://doi.org/10.1007/978-3-030-49020-1_3

Lytras, M.D., Visvizi, A. (2021) Information management as a dual-purpose process in the smart city: Collecting, managing and utilizing information. *International Journal of Information Management*. https://doi.org/10.1016/j.ijinfomgt.2020.102224

Malik, R., Visvizi, A., Skrzek-Lubasińska, M. (2021) The gig economy: Current issues, the debate, and the new avenues of research. *Sustainability*, 13, 5023. https://doi.org/10.3390/su13095023

Micek, G., Gwozdz, K., Kocaj, A., Sobala-Gwozdz, A., Świgost-Kapocsi, A. (2022) The role of critical conjunctures in regional path creation: A study of Industry 4.0 in the Silesia region. *Regional Studies, Regional Science*, 9(1), 23–44. https://doi.org/10.1080/21681376.2021.2017337

Nguyen, H.T., Marques, P., Benneworth, P. (2022) Living labs: Challenging and changing the smart city power relations?, *Technological Forecasting and Social Change*, 183, 121866. https://doi.org/10.1016/j.techfore.2022.121866

Nijkamp, P.; Kourtit, K., Türk, U. (2022) Special issue on The city 2.0 - Smart people, places and planning, *Journal of Urban Management*, 11(2), 139–141, https://doi.org/10.1016/j.jum.2022.05.011

Odendaal, N. (2016) Smart city: Neoliberal discourse or urban development tool? In: Grugel, J., Hammett, D. (eds) *The Palgrave handbook of international development*. London: Palgrave Macmillan. https://doi.org/10.1057/978-1-137-42724-3_34

Pellicano, M., Calabrese, M., Loia, F., Maione, G. (2019) Value co-creation practices in smart city ecosystem. *Journal of Service Science and Management*, 12, 34–57. https://doi.org/10.4236/jssm.2019.121003

Ramírez-Gordillo, T., Mora, H., Maciá-Lillo, A., Amador, S., Gil, D. (2024) Human-centric solutions and AI in the smart city context: The industry 5.0 perspective. In: Visvizi, A., Troisi, O., Corvello, V. (eds) *Research and innovation forum 2023. RIIFORUM 2023. Springer proceedings in COMPLEXITY*. Cham: Springer. https://doi.org/10.1007/978-3-031-44721-1_16

Sadowski, J., Pasquale, F.A. (2015) The spectrum of control: A social theory of the smart city (August 31, 2015). *First Monday*, 20(7), July 2015, U of Maryland Legal Studies Research Paper No. 2015-26, Available at SSRN: https://ssrn.com/abstract=2653860

Schuh, G., Zeller, V., Hicking, J., Bernardy, A. (2019) Introducing a methodology for smartification of products in the manufacturing industry. *Procedia CIRP*, 81, 228–233, https://doi.org/10.1016/j.procir.2019.03.040

Trencher, G. (2019) Towards the smart city 2.0: Empirical evidence of using smartness as a tool for tackling social challenges, *Technological Forecasting and Social Change*, 142, 117–128. https://doi.org/10.1016/j.techfore.2018.07.033

Vanolo, A. (2014) Smartmentality: The smart city as disciplinary strategy. *Urban Studies*, 51(5), 883–898. https://doi.org/10.1177/0042098013494427

Visvizi, A., Lytras, M.D. (2019a) Reflecting on oikos and agora in smart cities context: concluding remarks. In: Visvizi, A., Lytras, M. (eds) *Smart cities: Issues and challenges: Mapping political, social and economic risks and threats*. Elsevier. ISBN: 9780128166390. https://www.elsevier.com/books/smart-cities-issues-and-challenges/lytras/978-0-12-816639-0

Visvizi, A., Lytras, M. (2018) Rescaling and refocusing smart cities research: From mega cities to smart villages. *Journal of Science and Technology Policy Management (JSTPM)*. https://doi.org/10.1108/JSTPM-02-2018-0020

Visvizi, A., Lytras, M.D. (eds) (2019b) *Smart cities: Issues and challenges: Mapping political, social and economic risks and threats.* Elsevier. https://www.elsevier.com/books/smart-cities-issues-and-challenges/lytras/978-0-12-816639-0

Visvizi, A., Perez del Hoyo, R. (eds) (2021) *Smart cities and the UN SDGs.* Elsevier. https://www.elsevier.com/books/smart-cities-and-the-un-sdgs/visvizi/978-0-323-85151-0

Visvizi, A., Troisi, O., Grimaldi, M., Kozłowski, K. (2023) Getting things right: Ontology and epistemology in smart cities research. In: Visvizi, A., Troisi, O., Grimaldi, M. (eds) *Research and innovation forum 2022. RIIFORUM 2022. Springer proceedings in complexity.* Cham: Springer. https://doi.org/10.1007/978-3-031-19560-0_14

Part I
Spatial aspects of smart cities' growth and development

2 Path dependence, lock-in, and non-linearity in the growth and development of smart cities

Hanna Godlewska-Majkowska

2.1 Introduction

The objective of this chapter is to explain the mechanisms that shape the spatial structure of smart cities in line with the theory of economic geography. Based on the assumption that models of the spatial structure of cities are not compatible with the reality of smartification of cities with long-standing traditions, as well as emerging cities, this chapter aims to demonstrate the extent to which the path-dependency theory and the concept of lock-in can be helpful in explaining the spatial systems that form in a smart city. The topic of smart city development falls outside the mainstream discussion of smart city development. Considering that nowadays up to 50% of the cost of manufacturing products and services depends on the location, it is important to indicate the directions of spatial organization of smart cities and seek management recommendations for creating spatial order. To this end, this chapter identifies the theoretical basis for designing the processes of creating core structures in smart cities and attempts to present them in strategic documents for smart cities considered the leaders in Europe. In-depth interviews were conducted with representatives of city administrations, and regional development strategies were analyzed, particularly with regard to digitalization and the development of smart cities.

In this chapter, we attempt to find answers to the key research questions regarding smart cities' spatial developments, and whether there are similarities in the mechanisms of the spatial development of smart cities. In search of answers, we want to use the theories of dissipative structures, path dependencies, diffusion of innovation, and the concept of lock-in, putting forward the thesis that spatial development resulting from the current development trajectory has an impact on the processes of spatial development of smart cities. Strategic documents of these cities should contain references to spatial planning models or spatial recommendations, which is why additional research questions were also posed, regarding:

RQ1: Whether and how urban development models are used in the design of spatial order in smart cities.

DOI: 10.1201/9781003415930-4

RQ2: Whether there are common spatial development recommendations for smart cities that are characterized by similarities in terms of development trajectories, advancement of smartification processes, or degree of lock-in phenomenon?

2.2   From dissipative structures to a new spatial order

In recent years, the development of smart cities has become an incentive for the development of many regions, a concept developed in countries with different models of urbanization processes under different spatial, economic, social, and technological conditions. Characteristically, smart cities 1.0, 2.0, 3.0, and 4.0 are emerging as spatial overlays of already existing settlement units, and thus settlement systems.

This process is not planned. It takes place chaotically, creating points, nests, and, over time, linear or planar forms in the space of a given city, and urban complexes along with their suburban zones. They have the nature of dissipative structures (Prigogine & Lefever, 1973), a term borrowed from biology that provides an explanation for the processes occurring in the economic space. Dissipative structures are thermodynamic systems that are not in equilibrium and spontaneously generate order through energy exchange with the external environment (Goldbeter, 2018). This nonlinear and open system, which is not in equilibrium, constantly exchanges matter and energy with the external environment. In the case of cities, the external environment is the complementary area of the central place. The development of the city is at the expense of its surroundings, for which it is a pole of growth, both a source of resources and a market area.

Dissipative structures occur in living systems that are open, subject to nonlinear evolution equations, and also function far from thermodynamic equilibrium. The main source of nonlinearity is the multiple feedback processes that have evolved at the cellular and supracellular levels to optimize the functioning and survival of biological systems (Goldbeter, 2018). By applying this line of thinking to the spatial evolution of cities, including smart cities, it can be interpreted as the effect of feedback loops in the form of economic processes that operate in two directions, for example, through inverted supply chains in subsequent iterations created by investments with high investment multipliers and as the seed of the phenomenon of succession of city-building functions in suburban areas as a result of the depletion of reserves of investment land in a part of urban zones.

Small-scale irregularities can be transformed into large-scale patterns by dissipative structures, when far from equilibrium. When the change of a certain parameter in the system reaches a certain threshold, the phenomenon of self-organization can be generated by the effect of internal fluctuations and mutations, whereby the system is spontaneously transformed from its original chaotic state to an ordered state in time, space, or function (Gong et al., 2019), which means that new structural elements with a similar spatial pattern appear in the space of cities. An example of this are the transformation directions of traditional urban zones, which were initially individual buildings with the latest solutions based on Industry 4.0, from which, under the conditions of available investment space or the possibility of restructuring a particular central district, modern shopping and service center with characteristic, dense multistory buildings emerge over time, making extensive use of digital solutions. This can be observed, for example, in the center of Warsaw, where the expansion of the metro network has led to a significant density of this type of building in the Śródmieście and Wola districts near the metro stations.

The evolution of the system cannot be predicted, as there is always more than one evolving, qualitatively different structure. The transition to a new pattern may be relatively "smooth" or a sudden leap into a new domain, depending on the characteristics of the system and the disturbances (Gallopín, 2020).

Dissipative structures in cities are characterized by the absence of a clear order and plan and are instead the result of a city's random development and expansion. This is due to the rapid growth of cities and a lack of coordinated spatial policies and planning. An example of such a dissipative structure may be an area with a large number of fragmented housing developments, with chaotically distributed stores and services, and without a coherent transportation infrastructure. Such structures can involve both the traditional urban fabric and emerging smart city elements.

In a more advanced form, smart cities are created in a way that was planned from the beginning, based on reserves of investment areas in a suburban part of a city, often with suburban zone status, for example, Songdo in the Incheon Economic Zone in South Korea. This theory suggests that cities are nonlinear and open systems, far from a state of equilibrium, evolving through the exchange of energy and matter with the outside world. At the same time, there are limits to the nonlinear nature of cities, as they already exist in proximity to other elements of the settlement network, which is the reason for the increasing overlap of chaotically emerging smart city spatial structures with already formed structures in settlement systems.

The disorder arises from the overlap of two spatial structures –the structure of a traditional city rooted in the settlement system of a particular region/country/group of countries, traditional structures developed as a result of long-term urbanization processes, and spatial arrangements that are initially a network and then only after some time become spatial structures of the emerging smart city. These new structures initially develop chaotically, regardless of spatial order, based on the reserves of investment land in the city itself or in its surrounding area, and encounter obstacles in the form of the inertia of previously developed spatial structures and the longevity of previously made infrastructure investments.

The development of more orderly smart city structures is facilitated by the development of cities along main communication routes, which also determine the course of information and communication technology (ICT), for example, fiber optic grid. In addition, the restructuring of decapitalized residential, industrial, or office fabric also creates opportunities for the introduction of modern solutions. In this way, older structures can be replaced by an evolving smart city. The theory of dissipative structures provides an explanation for the emergence of smart cities as a spatial process. Based on the various manifestations of the influence of dissipative structures on the spatial development of a smart city, it is necessary to investigate whether:

- spatial structures in a smart city evolve at different scales, from microscopic elements to entire neighborhoods and cities, and, at different rates, depending on the specific factors affecting a given area;
- dissipative structures are created through an iterative process of adaptation and evolution that consists of constant adjustment to changing conditions;
- spatial structures are created by smart cities through self-organization of a network of nodes, leading to the emergence of hierarchical structural systems;
- spatial structures in a smart city are formed by the flow of information, knowledge, and data between different network nodes, leading to the emergence of specific patterns and directions of development.

From the development of smart city 1.0, 2.0, 3.0, and 4.0, we know that the formation of spatial structures in a smart city depends on many factors, such as urban policy, spatial planning, human behavior or the influence of technology, as well as existing development paths and related existing forms of spatial development.

Dissipative structures are formed based on the self-organization of spatial structures that openly spill out of administrative boundaries. The spatial scope of impulses that change spatial patterns is often caused by pre-existing structures, with long-term effects caused by the inertia of spatial structures. However, this phenomenon depends on the specific characteristics of a particular city's economic area and the characteristics and spatial extent of its sphere of influence.

2.3   Dependence on the path and lock-in as a contemporary determinant in the spatial development of smart cities

The explanation of the mechanism of self-organization of smart cities based on dissipative structures can be strengthened by using the theory of path dependence. This theory was shaped on the basis of microeconomics in the 1980s (Arthur, 1989; David, 1985) as a proposal to explain the processes of technology adaptation and industrial evolution, and especially why the free market mechanism in special circumstances leads to the spread of inferior products, despite the fact that there are alternative goods that satisfy the same need. Attention was drawn to the fact that suboptimal or inefficient technologies can be blocked as industry standards.

Researchers analyzed the reasons why the success of a given solution in production is often the result of an advantage gained at an early stage of development of a given branch of production. It turned out that the source of this advantage can be individual, seemingly unimportant events that cause a cumulative effect. This type of mechanism is associated with a virtuous circle in which each selection of a product based on a particular technology increases the likelihood that it will be chosen again. Standardization occurs as a result, which ultimately leads to the dominance of one solution and blocking other, alternative ways of development of a given industry.

This phenomenon is conditioned by a positive feedback loop, allowing for the cumulative effects of previous choices and the introduction of subsequent product units at a lower cost, which in turn leads to a dominant position on the market. These results show that initial decisions and events in the development process of an industry can be crucial for the formation of its future trajectories.

The further development of technology is impacted by previous decisions and technological choices, shaping competitive advantages and entry barriers for new solutions (Stack & Gartland, 2003). This theory has been used in mesoeconomic studies, which demonstrated that the ossification of spatial structures and the long-term effects of pre-existing regional specializations tend to continue existing developmental trajectories, even if there are significantly more profitable options for regional development at a given time.

Processes characterized by pathological dependence are inextricably linked to their previous course, which means that their development trajectory is sensitive to their own history. In the case of these processes, it is necessary to refer to the history of their course in order to fully understand their nature and dynamics. Proponents of this approach often use the term "history matters," meaning that history is important to understand and explain these processes. In the case of some processes, their explanation is impossible without referring to their past and their development over time. Such sensitivity to history stems from the fact that past choices and decisions have an impact on the direction of the process and its possible alternatives in the future, which is why an analysis of the history of the course of processes is crucial to fully understand their nature and predict their future trajectories (Domański, 2001; Szmigiel-Rawska, 2014).

In a broad sense, the concept of path dependence involves explaining the relationship between key past decisions and current and future states in a given chain of events (Sukiennik, 2017). In a narrower (otherwise formal) sense, researchers deal with identifying the initial conditions that gave a particular process its direction of development. To begin with, the elements responsible for the formation of the dependence path are identified, since the first step in the cause–effect process is often due to unforeseen or insignificant events, and these events usually have deterministic properties that reinforce a particular development path. Additionally, the beginning of the path that initiates the process of decision-making is not always rational, but sometimes the result of chance (Szmigiel-Rawska, 2014). Such unforeseen events could be individual decisions or large random events (natural disasters, wars, stock, and market collapses), which at the same time contradict the theoretical results predicted so far within a certain model (explanatory pattern).

Entering a different development path is associated with the need to make changes to the current spatial development of a particular city, leading to the continuation of the current strategic economic profile. Based on the lock-in concept, it can be concluded that the initial events in the development process are of key importance, as they can influence the formation of certain development paths and lead to the blocking of alternative paths. The term lock-in refers to a situation in which a product, behavior, or solution is constantly repeated and preferred by consumers. This repetitiveness results from the desire to reduce the costs associated with functioning in the marketplace and making alternative choices. The lock-in phenomenon is directly related to the initial events in the development process, which can affect consumer preferences and decisions and lead to the consolidation of certain development paths, meaning that the development of an area for new investment has already been completed.

A change in the use of a particular area is unprofitable as long as there are growth reserves, while closing off to other opportunities, also called lock-in, occurs when it is difficult to recoup the investments made to adapt a particular product or solution. In this case, the high cost of adapting physical and human capital, and increasing efficiency, may prevent the use of other, more beneficial alternatives. Clinging to one path leads to a limitation of options, because it blocks alternative solutions and disables taking advantage of their benefits, with being closed to other possibilities ultimately leading to difficulties in implementing change and innovation. In a situation where you are stuck in a particular sector or industry, it can be costly to choose a different solution, while switching to a different development path can be difficult. This effect also hinders the availability of alternatives, as choosing a dominant product or solution is a barrier to innovation and industry development.

The lock-in phenomenon is associated with increasing economies of scale and revenue accumulation. It occurs when the original solution becomes more profitable over time, limiting the ability to achieve alternative outcomes. The accumulation of revenues leads to positive and negative development cycles, the reversal of which is virtually impossible without outside intervention. The only way out of this situation is the intervention of an external force or a development shock that changes the structure of the system or radically reshapes the mutual relations between those involved.

Two types of model paths can be distinguished: self-reinforcing and reactionary. Self-reinforcing paths are cumulative, with a self-reinforcing sequence if it consists of recurring events. With self-reinforcing paths, the initial event stimulates development in that direction and is duplicated, making it difficult to leave the path after some time. Reactionary paths, on the other hand, have initial events that are not reinforced but transformed. Reactionary paths are temporally ordered sequences of cause and effect in which each event is both a reaction to earlier events and the cause of future events. The starting point of such pathways is the moment when several separate sequences are combined (Mahoney, 2000).

The concept of lock-in, also known as technology or network trap, is used in the context of urban development to explain processes where certain technologies, services, or solutions become dominant and their further development is hindered or blocked by existing structures, habits, resources, or relationships.

The lock-in concept assumes that an advantage is gained in the market by certain technologies or solutions, for example, through their early adoption, economies of scale, industry standards, or ecosystems around them. Insofar as such technologies or solutions gain acceptance, they create certain barriers to other alternatives that may be more efficient, sustainable, or innovative. Breaking the lock-in of technological systems and infrastructures is seen as critical for achieving more sustainable urban transformations.

In the case of urban development, the lock-in concept can affect various aspects, such as transportation infrastructure, energy systems, spatial planning, or economic patterns. For example, if a city invests in traditional transportation solutions such as highways and parking lots, it could create a technological trap hindering the subsequent development of more sustainable alternatives such as public transportation, bicycles, or pedestrian zones.

The concept of "lock-in" can also refer to socioeconomic patterns, such as the dominance of one economic sector, the concentration of resources in certain areas of the city, or the entrenchment of social inequalities. For example, if a city is heavily dependent on one economic sector, such as the oil industry, this could lead to a lock-in that would hinder the development of more sustainable and diversified economic sectors.

The lock-in concept can be used as a tool to understand urban development processes and identify barriers that impede innovation, sustainability, and the diversity of solutions. It can also help the formulation of urban policies and strategies that promote more flexible, diverse, and sustainable approaches to urban development.

2.4 Polarization of spatial development of smart cities (growth poles and center–periphery dichotomy)

With the development of smart cities, phenomena typical of the spatial diffusion of innovations occur, taking the form of polarization of urban space and consisting of the distinct formation of growth poles and peripheries in the newly emerging urban fabric with an accelerated course of the smartification process. Growth poles are large urban zones that are already technologically and infrastructurally well developed and have a well-educated society, and they often have access to cutting-edge technologies such as 5G networks, smart lighting, urban surveillance, and AI-driven transportation systems. They also have better financial resources and a more developed economy, allowing them to invest in smart city development.

The peripheries, on the other hand, are areas of the city that are technologically less developed, in terms of infrastructure and also socially. They may have less advanced transportation systems, lack telecommunications infrastructure, and have poorer housing and educational conditions. The periphery is often also more vulnerable to problems related to pollution and a lack of natural resources.

The pole–periphery dichotomy poses a challenge to the development of smart cities, as the concentration of investment and innovation in growth pole areas can further marginalize the periphery. It is therefore important that smart city projects are designed to include and develop all areas of the city, not just the most developed. The polarization of the spatial development of smart cities means that the development of advanced technologies is concentrated in selected parts of the city, while other areas lag behind the rapid implementation of digital solutions in the polar parts. In developed cities, where the concept of a smart city has already been implemented, the polarization of spatial development is particularly noticeable, a phenomenon that can lead to the emergence of the so-called "subcities" – urban zones with different levels of development and access to services. This is particularly evident in the spatial structure of cities.

The polarization of the spatial development of smart cities is caused by many factors that can be attributed to the activities of individual stakeholders. In the development phase of smart city 1.0, the decision of companies offering new technologies, for example, in the field of lighting of common areas, as well as the location of buildings with service, commercial, or industrial functions, is of great importance, including unequal access to modern technologies. Not all residents have the same level of Internet access or devices that enable the use of smart city services. In addition, there are differences in the quality of services, which can lead to the social exclusion of certain groups of residents. Polarization in this case is related to the attraction of new investments by local authorities through the creation of an offer of investment areas, and thus the uneven distribution of financial resources for the development of smart cities, which is characteristic of the development of smart cities 2.0. For example, if science and technology parks, local industrialization, or economic activation zones are created, then such initiatives can also lead to the creation of smart city growth poles based on the creation or modernization of the city's economic base. Such initiatives are also triggered by the creation of new locations for special economic zones, justified by cases such as the creation of the Songdo Smart City in South Korea.

The polarization of the spatial development of smart cities is also influenced by the public demand for public services and buildings with the highest standards. With a growing sense of security and the desire to protect themselves from theft or other disruptions to their sense of security, wealthy neighborhoods are becoming more heavily equipped with the latest surveillance solutions, and housing and local retail and service provision are increasingly using the latest Internet of Things solutions. It is also worth paying attention to the overcrowding of neighborhoods inhabited by immigrants, people without a stable income, or those living in big cities only temporarily – magnets for mass migration, both economically and politically.

In such areas, there is no demand for cutting-edge digital solutions, as the inhabitants of such neighborhoods often struggle with digital exclusion or even show frustration due to a much lower standard of living compared to gated neighborhoods inhabited by the economic and political elite of a particular urban zone. This means that residents of these urban zones also have less access to the benefits of smart city solutions, such as transportation services, pollution monitoring, or the availability of city information.

Crowded neighborhoods also have no room for investments typical of smart cities, hence, the mosaic characteristic of large cities in the creation of smart city spaces, where old traditional architecture is mixed with modern "oases" to form smart city islands.

The spatial development of the city is also uneven due to differences in the cost-effectiveness of deploying individual digital solutions for public services. It is inevitable that densely populated areas will attract corresponding upfront investments from city governments due to the number of users compared to sparsely populated areas or green spaces.

The polarization of the spatial development of smart cities is a phenomenon that can lead to social inequalities and social exclusion, and it is therefore important that investments in smart cities take into account the needs of all residents, not only those of the wealthier neighborhoods.

The mechanism of this phenomenon is complex and depends on many factors, such as location, government policy, availability of funding, innovation by entrepreneurs, and acceptance and social engagement. One of the key factors is also the ability to attract investments and talent, as well as a favorable environment for start-ups and entrepreneurship. This requires well-thought-out policies in the field of shaping the conditions for sustainable development, taking into account the spatial order, which is also expressed in well-thought-out decisions on the chosen models of spatial development of smart cities. If we look at the example of Warsaw, characteristic spatial regularities can be observed. The highest density of measuring points for vehicle traffic can be found in the central part of the city and along the communication lines of supralocal importance (Figure 2.1).

The higher density of noise measurement points in the left bank of Warsaw is also interesting to note and is related to the fact that the western part of Warsaw and its suburban areas are more industrialized than the districts of the right bank and the eastern suburban part. It is also worth paying attention to the not always visible relationship between the density of these points and traffic intensity. An example of this is the isolated points in outlying areas that are located away from traffic routes or have high population densities, where car traffic is relatively low, indicating the diverse involvement of local authorities in the creation of this type of infrastructure.

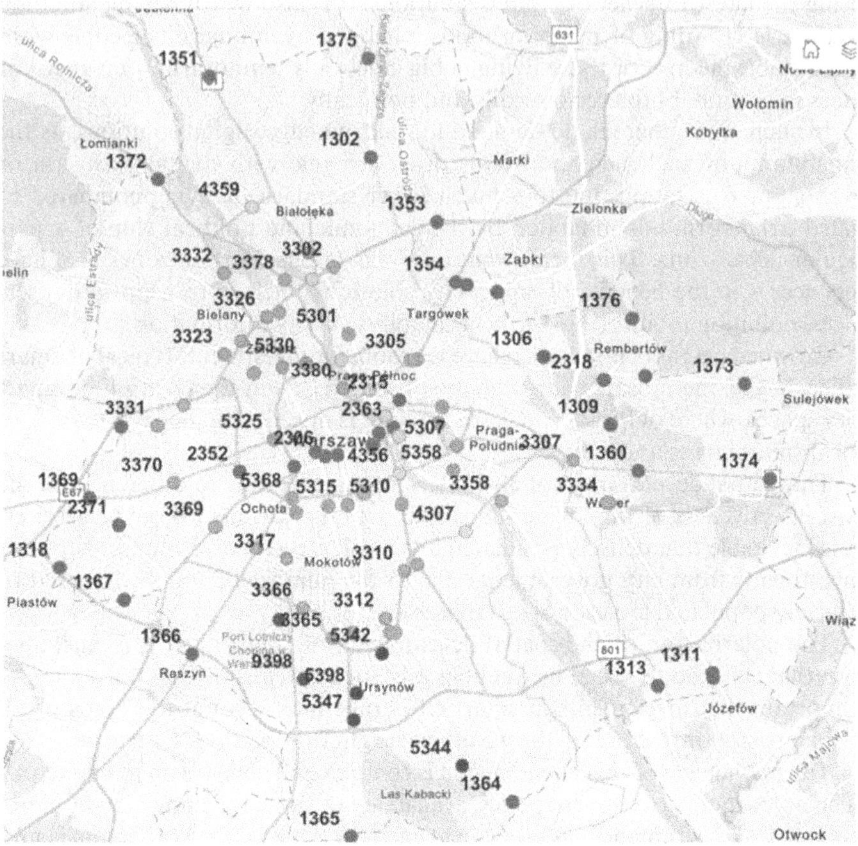

Figure 2.1 Distribution of traffic measurement points in Warsaw, as of April 30, 2023.

Source: https://www.google.com/maps/d/u/0/viewer?mid=11x9Leh1rziFJLd-JSrEDG574uNpGiO Lm&ll=51.95341819154865%2C20.897021689795288&z=12

A similar phenomenon can be seen when analyzing the arrangement of urban lighting based on the latest generation of LED lamps. Although Warsaw is largely equipped with a network of smart lanterns, some parts of the city's residential neighborhoods are better lit than traffic arteries. Grochowska Street is a good example of this – with a small number of lamps, it is one of the busiest streets on the right side of Warsaw and parallel to Łukowska Street, a street that mainly serves the needs of residents of the Witolin housing estate.

2.5   Spatial models – Types and choice dilemmas

The development of smart cities is inextricably linked to the use of digital technologies and the smartification of the economy and is therefore a stage in the development of cities that often have a centuries-old tradition of development.

In such cases, when it comes to restructuring an existing city, revitalizing a city, or expanding it, there is a very large influence of historical factors and the previous course of urban development.

The urban landscape is sustainably developed, giving priority to the renovation and development of existing environ mentally friendly buildings while preserving the particular identity of each area (Locurcio et al., 2023). Smart cities are thus created as a smart, specific layer, as an overlay on an already existing city, based on modernization and replacement investments that promote the use of modern digital solutions throughout the city. Additionally, smart cities are created based on new investments in areas that form the investment areas of a specific city, or on land connected to a specific city, along with its spatial boundaries.

The spatial limitation in the development of smart cities is infrastructural and is caused by differences in access to the Internet and the intensity of digital exclusion of the residents of different parts of the city. These differences are historically conditioned, thus exhibiting path dependency, and are subject to the lock-in phenomenon previously described.

Historically conditioned densification processes from construction using smart solutions take place along communication paths, close to junctions. An example of such a phenomenon is the development of Helsinki, where new urban spaces are created along railroad lines and stations (City of Helsinki, 2021, 36).

A characteristic feature of this development is infilling. Infilling of investment space reserves is associated with the use of a historically shaped transport network, and, in this sense, historical conditions, path dependency, and the phenomenon of lock-in manifest in the development of smart cities as another layer overlaying the spatial development of the city. With the development of suburbanization and new urban districts, smart city districts are created from scratch. Helsinki's development strategy, however, does not mention the development of Itäkeskus as a smart city.

Another reference to the model of core structures in a smart city can be found in the smart city development strategy (City of Vienna, 2022, 72). The smart city aims to create multifunctional sustainable economic structures in the city by creating mixed-use districts. Innovative production facilities must be located near high-quality services and research and development facilities, as intelligent solutions under one roof play an important role. Modern development concepts offer residential and office space, commercial and service areas, and cultural and leisure facilities under one roof.

Further, the focus is on developing existing neighborhoods – both in existing buildings and in new districts – which provide space for commercial and noncommercial, social, and community use. The Neighborhood Stimulus Campaign aims to revitalize, strengthen, and improve local neighborhoods and main streets and promote modern forms of work by mobilizing vacant spaces and creating neighborhood-serving office space. There is a visible reference to the concept of the spatial structure of a city based on the sequence of city-forming functions – the polycentric (multicentric) model (Harris & Ullman, 1945). According to this model, a city can form multiple urban zones by creating focal points in its peripheral areas. However, while in the polycentric model, it was clear to preserve the monofunctional character of the use of real estate or city sectors, in the new approach of a smart city model, there is a tendency to develop the multifunctional character of the use of a particular real estate in order to shorten the daily mobility routes of residents as much as possible.

The compact city model aims to stop urban sprawl, support the revitalization and development of neglected areas, prioritize public transit, cycling, and pedestrians, and limit automobile traffic. This is a response to the development of dispersed cities (Mierzejewska, 2015). The benefits of a compact city include land saving due to high building density, economical communication (independence from cars), cheaper technical infrastructure, orderly and well-kept public spaces, and availability of social services (Heubeck, 2008). The issue of global warming is also reflected in Vienna's smart city development strategy. Projections and simulations based on historical data suggest that solutions to prevent the effects of global warming should be introduced into the smart city space, focusing on supporting particularly vulnerable populations.

Densely built-up inner-city areas are disproportionately affected by the "urban heat island effect." Fresh air corridors and cold air flows should be taken into account at the urban planning stage, in order to create an interconnected network of high-quality open and green spaces. New investments should not create additional heat islands but actually improve the urban microclimate. Therefore, interior development places significant emphasis on cooling and greening agents, water, and shade – which is much more efficient and environmentally friendly than air conditioning.

The concept of sponge cities, a term first introduced by an Indian author in 2005 (Van Rooijen et al., 2005), is noteworthy. The concept suggests that the basis for stormwater management are surfaces that allow water to naturally run off or evaporate – cooling the air while relieving pressure on the sewer system, while a porous retention layer beneath the surface of a street or sidewalk provides water to trees along the street.

The spatial structure of a smart city should also include green belts that allow air to circulate and prevent the formation of urban heat islands. The ecosystems of the green belt that surrounds Vienna and the green corridors that cross it are the "green lungs" of the city, and the cool air they produce has a major impact on the urban microclimate. The size and quality of these ecosystems must therefore be protected in the future and remain barrier-free and easily accessible by foot, bicycle, and public transport. The Vienna experience is a solution of great value for the development of smart cities in different regions of the world and something that can also be applied without interlocking constraints in new smart cities created from scratch, especially in previously undeveloped areas, areas reclaimed from the sea, or areas previously used mainly for agriculture or forestry.

In the literature on this topic, there are model approaches to the spatial development of a city with a more universal character, which can therefore also be used for spatial planning in cities that are becoming smart cities. These models include the concepts of a 15-minute city and a linear city.

The 15-minute city model (Moreno et al., 2021) is based on the assumption that a high quality of life can be maintained for residents by limiting mobility time to 15 minutes to move between key points in their daily cycle, such as home, work, and school, allowing residents to meet their basic needs in a short amount of time by walking or cycling. This model puts focus on the importance of increasing the density of urban zones through multifunctionality and improving the quality of life.

Implementation of the 15-minute city concept is based on four main principles: diversity (in buildings and culture), digitalization (using the smart city concept), proximity (in the temporal and spatial sense), and density (number of people per square meter). In line with this model, a monofunctional urban space can be transformed into a multifunctional space that enables the simultaneous implementation of these four components.

The concept of the 15-minute city first appeared in Paris and then quickly spread to other cities around the world. It is currently being applied in many metropolitan areas around the world and in Europe, such as Bogota, Melbourne, Detroit, Portland, Ottawa, Shanghai, Barcelona, London, Vienna, and Milan. An integrated urban transport network and green corridors have been introduced in these cities, enabling the implementation of the 15-minute city concept and improving the quality of life of residents.

The linear city model is built on universal space segments that ensure scalability while minimizing the duration of the daily cycle of the inhabitant and is created based on certain repetitive spatial development patterns, such as those associated with the idea of 20-minute access to basic public services. One attempt to implement this concept is The Line, a city being built in Saudi Arabia. The project involves the construction of a 170-kilometer strip of buildings that will run in a straight line from the city of Neom to the east in Saudi Arabia. The city will be made up of "modules" and will connect the Red Sea coast with the northwestern regions of the country. There will be no cars or roads in the linear city of the future. Instead, it will be full of green parks, open spaces, and ultra-fast public transportation. According to the plan, one million people from all over the world will settle in The Line. Importantly, all residents will live within walking distance of key amenities, and journeys between hubs should take no more than 20 minutes. Due to the many risks associated with futuristic projects, efforts in the case of smart cities are aimed at thoughtful, incremental improvements to existing cities rather than attempts to design entirely new cities from scratch.

2.6 Conclusions

The aim of this chapter is to explain the mechanisms that shape spatial structure. Based on the theories presented to support the analysis of the core aspect of smart city development, several conclusions can be drawn.

First, applying the theory of path dependence can be helpful as a conceptually important basis for finding spatial patterns in urban development. The historically shaped spatial patterns of cities are reflected in the development of smart cities based on investments through the densification of existing residential, commercial, and service buildings. Spatial systems in emerging smart cities will be of an overlapping nesting structure nature (sealing by individual buildings and structures using ICT), linear as a result of the creation of axes of smart city development based on the existing and expanded transport system, and areal based on new or restructured districts using smartification of production, service, and housing.

Path dependency can also be used to explain how new districts are created, often perceived as cities in their own right, such as Songdo, where the restructuring of a particular area is combined with the creation of economic privilege zones. These privileges are often linked to the desire to increase the investment attractiveness of a particular urban zone and open it up to the development of solutions typical of an emerging smart economy.

Based on the selected smart cities under analysis, similarities in the mechanisms of smart city spatial development can be observed. The most characteristic mechanism is based on the diffusion of innovations, which take place in a hierarchical manner, that is, from the center of a given urban zone to its periphery. Such tendencies are visible, for example, in the analysis of selected elements of an evolving smart city, such as the digital analysis of traffic intensity or the spread of smart city lighting systems used. Case studies have shown that the process of diffusion of infrastructure innovations typical of smart cities (e.g., based on the Internet of Things) depends on the associated savings, making the primary beneficiaries of this solution the central districts, which are the most densely populated and have a high density of buildings with economic functions and tourist attractions. Thus, the diffusion of innovation, which is an expression of the creation of a smart city based on already existing cities, is centrifugal from the center to the periphery.

An obstacle to the diffusion of innovation leading to smart cities is the cost associated with changing the current use of urban space, which is due to the lock-in phenomenon. Once specialization of the economic base is in place, it is difficult to transfer it to the specialization of the entire economic ecosystem of a given urban area, and the lack of free investment space causes this type of investment to be pushed to the peripheral parts of a given urban zone. How this happens can be explained by the theory of dissipative structures.

Self-organizing structures emerge in smart cities, thanks to the effect of similar driving forces in individual cities that trigger the self-organization of the economic space. Recurring patterns in the spatial structure of cities are caused by increasing emphasis put on improving the quality of life in the city, preventing the formation of urban heat islands, and achieving the lowest possible costs for the implementation of smart infrastructure solutions. A spatial tendency toward multifunctional densification of urban development was therefore observed in all the cities analyzed, aimed at reducing the daily mobility of residents or commuters in the context of circular migration. In addition, there is a discernible desire to create districts with higher building standards in terms of security of advanced ICT technologies, which favors the creation of similar specialized areas for technology parks or other such economic activation zones for industry and high-tech services. References to models for spatial structures or spatial recommendations should be included in the strategic documents of cities.

Thus, the self-organization of smart city space leads to densification of the central parts of the city, filling them with elements typical of a sponge city with a large proportion of green spaces. There is also a growing tendency to create specialized suburban areas that are planned from the outset as multifunctional areas to maintain spatial order.

The development of cities is not sufficiently strengthened by the creation of new theoretical models. The most common models of an urban spatial structure were created during industrialization and are therefore not suitable for the current trends of de-congestion of economic activity and the business models based on remote work. The only trend that deserves attention is the sponge city model, which responds to the needs of change in the face of global warming, as well as to the competition among large cities for talented residents, which requires them to compete by offering competitive living conditions based on the principle of longevity.

The challenges and mechanisms described above are not adequately addressed in smart city development strategies, where there are usually recommendations for complementary investments based on existing development, but there is a lack of authentic vision. The only good practice identified in this area is the smart city development strategy for the Netherlands. Therefore, as a recommendation from the conducted analyses, the following theory can be formulated: it is not enough to be guided by the profitability of individual investments when developing smart city infrastructures. Rather, it requires a vision and a clear definition of what the city is in order to compete with other smart cities over a period of several decades. Despite the high variability of competitive conditions, investments in the city entail a change in the way the city develops once they are finished, which is the idea behind the self-organization of dissipative structures and the lock-in mechanism.

Acknowledgments

The research was conducted at the SGH Warsaw School of Economics in the College of Business Administration under the direction of H. Godlewska-Majkowska and covered the following cities: Barcelona, Duesseldorf, Gdynia, Gothenburg, Hanover, Katowice, Kiel, Leeds, Łódź, Rotterdam, Lisbon, Verona, Vienna, Warsaw, and Wrocław. Research presented in this chapter constitutes a part of the implementation of the following grant: SGH Warsaw School of Economics, Statutory Research Grant, titled: Smart Business – Smart Regions – Smart Society: On the Way to a New Paradigm (1.1. KNOP/ S22), led by Hanna Godlewska-Majkowska.

References

Arthur, B. (1989) Competing Technologies, Increasing Returns, and Lock-in by Historical Small Events. *Economic Journal*, 99(394), 116–131.

City of Helsinki (2021) A place of growth. Helsinki City Strategy 2021–2025. https://www.hel.fi/static/kanslia/Julkaisut/2021/helsinki-city-strategy-2021-2025.pdf (accessed on 10.09.2022).

City of Vienna (2022) Smart Climate City Strategy Vienna. Our Way to Becoming a Model Climate City. Vienna's Strategy for Sustainable Development. 2022. https://smartcity.wien.gv.at/wp-content/uploads/sites/3/2022/05/scwr_klima_2022_web-EN.pdf

David, P. A. (1985) Clio and the Economics of QWERTY. *The American Economic Review*, 75(2), 332–337. http://www.jstor.org/stable/1805621

Domański, B. (2001) Kapitał zagraniczny w przestrzeni Polski. Prawidłowości rozmieszczenia, uwarunkowania i skutki, Instytut Geografii i Gospodarki Przestrzennej Uniwersytetu Jagiellońskiego, Kraków.

Gallopín, G. C. (2020) Cities, Sustainability, and Complex Dissipative Systems. A Perspective. *Frontiers in Sustainable Cities*, 2, 523491. doi: 10.3389/frsc.2020.523491

Goldbeter, A. (2018) Dissipative Structures in Biological Systems: Bistability, Oscillations, Spatial Patterns and Waves. *Philosophical Transactions of the Royal Society A: Mathematical, Physical and Engineering Sciences*, 376, 20170376. doi: 10.1098/rsta.2017.0376

Gong, Q., Chen, M., Zhao, X., Ji, Z. (2019) Sustainable Urban Development System Measurement Based on Dissipative Structure Theory, The Grey Entropy Method and Coupling Theory: A Case Study in Chengdu, China. *Sustainability*, 11, 293. doi: 10.3390/su11010293

Harris, C. D., Ullman, E. L. (1945-01-01). "The Nature of Cities". *The Annals of the American Academy of Political and Social Science*, 242, 7–17.

Heubeck, S. (2008), Competitive Sprawl. *Economic Theory*, 39 (3), 443–460.

Locurcio, M., Tajani, F., Anelli, D. (2023) Sustainable Urban Planning Models for New Smart Cities and Effective Management of Land Take Dynamics. *Land*, 12(3), 621. doi: 10.3390/land12030621

Mahoney, J. (2000) Path Dependence in Historical Sociology. *Theory and Society*, 29, 510–512.

Mierzejewska, L. (2015), Miasto zwarte, rozproszone, zrównoważone. *Studia Miejskie*, 19, 9–22.

Moreno, Carlos, Allam, Zaheer, Chabaud, Didier, Gall, Catherine, Pratlong, Florent. 2021. "Introducing the "15-Minute City": Sustainability, Resilience and Place Identity in Future Post-Pandemic Cities" *Smart Cities* 4(1), 93–111. https://doi.org/10.3390/smartcities4010006

Prigogine, I., Lefever, R. (1973) *Theory of Dissipative Structures*. Springer: Berlin, Germany.

Stack, M., Gartland, M. P. (2003) Path Creation, Path Dependency, and Alternative Theories of the Firm. *Journal of Economic Issues*, 37(2), 487–494. http://www.jstor.org/stable/4227913. Accessed 23 Apr. 2023.

Sukiennik, J. (2017) Path dependence – proces przekształceń instytucjonalnych, "Prace Naukowe Uniwersytetu Ekonomicznego we Wrocławiu" 2017, nr 493, s. 164–165.

Szmigiel-Rawska, K. (2014) Koncepcja zależności od ścieżki jako narzędzie wyjaśniania w badaniach ekonomicznej geografii politycznej, "Prace i Studia Geograficzne" 2014, t. 54, s. 149–161.

Van Rooijen, D. J., Turral, H., Biggs, T. W. (2005) Sponge city Water Balance of Mega-City Water Use and Wastewater Use in Hyderabad, India. *Irrigation Drainage*, 54(2005), S81–S91.

3 The smart city and its contexts

A focus on smart villages and smart territories

Małgorzata Dziembała, Radosław Malik, and Anna Visvizi

3.1 Introduction

Socioeconomic development within a country is not uniformly distributed as certain regions experience substantial economic growth, while others manifest minimal or stagnant growth rates (Döring and Schnellenbach, 2006). The economic development of territorial units is the result of complex and diverse factors (Qiang and Jian, 2020). The factors that stimulate development include, among others, a modern and diversified economic structure, infrastructure in terms of its quality and innovativeness, high quality of human capital, and quality of governance (Urbano et al., 2019). Typically, growth has been concentrated in urban areas, while rural areas have experienced a number of difficulties and problems. These would result in lower growth rates (Li et al., 2019). Consequently, the dichotomy between urban and rural developments has emerged as a key question in the domains of economic geography, regional development, and development studies (Porru et al., 2020; Mettler and Brown, 2022; Luo et al., 2022).

The distinction between urban and rural areas presents significant challenges for economic policies aimed at addressing societal challenges, curtailing economic and social polarization, and fostering greater cohesion. Cohesion can be conceptualized as the extent, to which economic, social, and territorial disparities are politically acceptable (Medeiros et al., 2023). While cohesion, and its local, regional, national, and international levels, has been the subject of political and academic debates for a great number of years now, recently, it has returned to the EU level policy discussions (Lillemets et al., 2022). In response to these deliberations, a series of programs and measures were launched, designed to ameliorate disparities, particularly those associated with the urban–rural divide (Di Caro and Fratesi, 2022).

Advances in ICT-based tools can be perceived as a strategic instrument to bolster cohesion, given the pivotal role ICT plays in fostering the growth of services that transcend traditional geographical constraints (Cristofoletti et al., 2023). Literature suggests that smart cities establish an optimal environment for the proliferation of such services, and their evolution could markedly affect the urban–rural dichotomy (Gong and Shan, 2023; Kędra et al., 2023). Additionally, by blurring the demarcation between the urban and the rural, the

DOI: 10.1201/9781003415930-5

underlying dynamics might facilitate the genesis of virtual spaces that integrate attributes of both urban and rural areas, thereby forming smart territories underpinned by service ecosystems (Gutierrez-Velez et al., 2022; Visvizi and Lytras, 2018). Consequently, it becomes imperative to investigate certain research questions to enhance comprehension regarding the advancement of smart cities, particularly in the context of the increasing prevalence of smart services and their impact on the urban–rural division.

Q1. Can smart cities serve as a bridge between the urban and rural divide?
Q2. What are the primary dimensions of the disparity between urban and rural regions within the framework of the EU's cohesion strategy?
Q3. To what extent and how does the intrusion of ICT-enhanced tools and applications in the city space impact a city's relationship with its broader geographical context?

To address these questions, the argument in this chapter is structured as follows. The next section offers an insight into the existing literature to suggest how to conceive of the smart city, especially in relation to the urban–rural divide. The following section explores the urban–rural dichotomy, stressing the challenges of cohesion in the EU context. Then, a conceptualization of the smart city and its contexts is elaborated. The idea here is to capture the dynamics of the smart city, its contexts, and the underlying mechanisms that condition the smart city's interactions with external environments. The services' ecosystem approach plays a key role in this framework. Discussion and conclusions follow.

3.2 Literature review

Historically, cities have been centers of trade and commerce, facilitating the exchange of goods and ideas and traditionally played a pivotal role in the growth and evolution of economies acting as hubs of innovation, commerce, and culture. In the post-industrial era, the services sector has become predominant in cities and delineating the economic geography of cities their role in providing non-tradable services is emphasized. Urban service ecosystems encompass various domains, including, predominantly: financial services, educational and health services, information and technology services, entertainment, and cultural services (Haase et al., 2014). The growing importance of cities in their provision of increasingly sophisticated services contributed to the development of the concept of "global cities," highlighting their importance as centers of innovation and specialized, information-based services that are not provided just to their immediate hinterland but serve a global clientele. This emphasizes the gravitational pull cities exert in attracting both human and financial capital, which subsequently drives their role as primary contributors to economic growth (Chakravarty et al., 2021).

The center–periphery paradigm has been adapted to elucidate the interactions between metropolitan hubs and their adjoining rural regions (Sawyer et al., 2021). Within this framework, urban centers emerge as the "core," orchestrating a convergence of wealth, vital resources, and intellectual capital. The agglomeration of economic undertakings in these centers engenders pronounced economies of scale, catalyzing both innovation and sustained growth. Conversely, rural areas, representative of the "periphery," are predominantly anchored in agricultural or resource-centric endeavors and thus confront challenges in diversifying their economic profiles (Zöllner et al., 2023). The concept of "trickle-down" economics gains prominence in this narrative, suggesting that in time, growth in urban areas stimulates growth in peripheral areas (Vergara and Salazar, 2021). Furthermore, urban centers, through diverse services available therein, e.g., financial, educational, and healthcare sectors (cf. Visvizi et al., 2017), exert a pivotal influence over the broader economic terrain, permeating even into the periphery. Nonetheless, this dynamic is not devoid of challenges. The inexorable pull of urbanization, while seemingly advantageous for peripheral regions by spurring demand, may inadvertently precipitate a drain of talent and critical resources. Such phenomena can amplify existing disparities, deepening the urban–rural divide (Sano et al., 2020).

The evolution of smart cities is a multifaceted process wherein smart services are collaboratively fashioned by diverse stakeholders, encompassing economic, technological, social, and cultural dimensions (Lytras and Visvizi, 2018). Recent studies have elucidated that smart services, underpinned by the pivotal role of ICT technologies, are integral to the overarching concept of a smart city (Malik et al., 2022). From this perspective, smart services act as fundamental mechanisms propelling the development of smart cities, establishing intricate relationships between citizens and the services they utilize (Visvizi and Lytras, 2018). Given the proliferation and the value generation potential of smart services, there is a burgeoning interest in pioneering frameworks such as smart service ecosystem analysis (Troisi et al., 2019).

The smart service ecosystem, in the context of smart cities, can be viewed as an integrated and interconnected set of digital services and solutions designed to optimize urban living. This ecosystem encompasses a broad array of components, including advanced sensors, data analytics platforms, network infrastructure, and application interfaces, all orchestrated human–technology interactions (Kashef et al., 2021). Central to this are the Internet of Things (IoT) devices, which constantly gather data, and the underlying cloud infrastructure that analyzes and acts upon this data in real time. Additionally, stakeholder participation, from both public and private sectors, plays a pivotal role in shaping and refining these services (Wirtz and Müller, 2023). When effectively implemented, the smart service ecosystem fosters enhanced urban efficiency, sustainability, and livability. Ultimately, it aims to engender a seamless urban experience, marked by reduced resource consumption, improved public services, and elevated citizen well-being (Palumbo et al., 2021).

Smart cities research should inherently reflect the experiences of individuals residing in both rural and urban environments, and it ought to align with and influence policy design and decision-making (Visvizi and Lytras, 2018). To truly advance global sustainability, it is imperative to comprehend the intricate interplay between rural and urban processes and their subsequent impact on sustainability outcomes. Historically, the examination of rural–urban relationships has been predominantly urban centric, with urbanization perceived as the primary determinant of sustainability results. Researchers have advocated for a paradigmatic shift in this understanding, urging scholars to pivot from an exclusively urban viewpoint and accentuate the influential role of rural processes, practices, and locales in driving rural–urban sustainability (Gutierrez-Velez et al., 2022). Such a nuanced perspective has the potential to catalyze the development of innovative infrastructures in smart territories, fostering a harmonious integration between rural and urban landscapes.

Smart territories encompass a broader scope beyond the conventional smart city concept, representing regions that integrate digital technology, data, and innovative strategies across urban, suburban, and rural areas (Garcia-Ayllon and Miralles, 2015). Characteristic features of smart territories include holistic urban planning, optimized resource management, and enhanced connectivity through integrated digital infrastructure. These territories not only prioritize technologically advanced services but also emphasize sustainability, citizen participation, and regional inclusiveness (Navío-Marco et al., 2020). The outcome of adopting a smart territory approach is the creation of resilient, efficient, and sustainable environments. Moreover, it fosters greater collaboration among different regions, ensuring that advancements and benefits aren't confined to densely populated urban centers alone and can exert their influence in a global scale supporting the virtual space of smart service creation and exchange.

3.3   Urban and rural areas and the question of cohesion in the EU

The development of regions within the EU encounters a myriad of challenges juxtaposed with substantial opportunities emerging from digital transformation. The nuances and features of these challenges and opportunities diverge considerably across urban locales. To effectively navigate these challenges and harness the potential of emerging opportunities, profound structural alterations are imperative. An unintended consequence of these shifts might manifest in cohesion-related risks. Foreseen demographic shifts will predominantly impact rural regions, marked by an escalating population aged 65 and above coupled with a diminishing workforce demographic. Consequently, such challenges could amplify the pre-existing disparities between rural and urban areas. These disparities currently span dimensions such as GDP and income, poverty and unemployment rates, educational attainment and skills, and levels of trust and social capital.

The territorial landscape of the EU exhibits significant regional heterogeneity, characterized by areas with robust GDP per capita juxtaposed against those that remain underdeveloped. Notably, these disparities are more pronounced along the axis of rurality versus urbanity. Predominantly, urban regions documented the highest GDP per capita, whereas intermediate and principally remote rural areas registered GDP per capita at merely 68% and 69% of the EU average, respectively (Table 3.1). Furthermore, when assessing the variances between towns, urban, and rural regions across EU member

Table 3.1 GDP per head by urban-rural regional typology in 2000–2019

GDP PER HEAD (PPS), EU27 = 100	2000	2010	2011	2012	2013	2014	2015	2016	2017	2018	2019
EU-27	100	100	100	100	100	100	100	100	100	100	100
Predominantly urban	130	128	128	127	126	126	126	125	125	125	125
Intermediate, close to city	89	88	89	89	89	89	90	89	89	89	89
Intermediate, remote	71	72	71	71	71	69	69	68	67	68	68
Predominantly rural, close to city	69	72	72	73	74	74	75	77	76	77	78
Predominantly rural, remote	70	70	70	69	69	70	70	69	69	69	69

Source: Figures and tables in Excel, Eighth Report on Economic, Social and Territorial Cohesion. https://ec.europa.eu/regionalpolicy/information-sources/cohesion-report-en, 15.04.23.

states, net income emerges as a pivotal metric (Table 3.2). The disparity in net income between cities and their rural counterparts averaged 3,971 euros across the EU-27, reaching peaks in Luxembourg and Ireland. Nonetheless, certain member states presented deviations from this trend. Empirical evidence from 2021 indicates that, on average, cities in EU member states boasted superior mean net incomes, with Belgium, Malta, the Netherlands, and Austria serving as notable exceptions. This specific indicator underscores the economic facet of regional discrepancies. Thus, an imperative emerges to address these inequities in pursuit of enhanced territorial cohesion.

An examination of unemployment rates across urban and rural sectors within the EU further underscores the pronounced divide between these regions. In 2021, the unemployment rate for individuals aged 15–74 in the EU-27, when analyzed by the degree of urbanization, elucidates the disparities from both economic and social standpoints (Table 3.3). The findings present nuanced outcomes concerning unemployment rates across different urbanization categories. Broadly, cities exhibited higher unemployment rates compared to rural regions. Nonetheless, certain countries deviated from this general trend. Unemployment serves as a significant determinant influencing the risk of poverty. The social dimension of the disparities between rural and urban areas is further epitomized by the "persons-at-risk-of-poverty" metric, as well as indicators of social exclusion (Table 3.4). Data from 2020 for the EU-27 reveals that individuals at risk of poverty or social exclusion constituted 22% in urban settings and 22.8% in rural contexts. On a broader scale, rural zones within EU Member States exhibited a marginally elevated rate of exclusion compared to their urban counterparts. In countries such as Romania and Bulgaria, the urban percentages stood at 31.3% and 23.7% respectively. Intriguingly, urban

Table 3.2 Mean equivalized net income by degree of urbanization in 2021 (in EUR)

EU member countries	Cities	Towns and urban areas	Rural areas	Difference between the value for city and the value for rural areas
EU–27 countries (from 2020)	22 241	21 325	18 270	3 971
Euro area – 19 countries (2015–2022)	24 488	23 680	22 394	2 094
Belgium	25 483	29 030	27 418	−1 935
Bulgaria	8 320	5 966	5 081	3 239
Czechia	13 295	11 337	11 623	1 672
Denmark	36 100	37 212	33 941	2 159
Germany	29 622	29 037	28 186	1 436
Estonia	15 009	13 380	13 703	1 306
Ireland	34 712	33 439	28 220	6 492
Greece	10 666	10 508	8 501	2 165
Spain	19 482	16 438	16 235	3 247
France	27 911	25 029	24 658	3 253
Croatia	10 173	8 705	7 859	2 314
Italy	21 462	20 016	18 561	2 901
Cyprus	21 487	17 515	16 201	5 286
Latvia	12 231	12 029	9 522	2 709
Lithuania	13 766	9 966	10 689	3 077
Luxembourg	55 509	44 860	48 822	6 687
Hungary	8 357	6 983	6 690	1 667
Malta	18 865	20 944	19 649	−784
Netherlands	30 270	31 161	30 858	−588
Austria	29 100	30 808	30 176	−1 076
Poland	10 522	9 201	7 861	2 661
Portugal	14 578	12 628	10 839	3 739
Romania	7 523	5 446	3 956	3 567
Slovenia	17 499	16 667	16 160	1 339
Slovakia	10 062	8 769	8 671	1 391
Finland	30 547	27 623	26 975	3 572
Sweden	29 265	27 045	25 558	3 707

Source: Own compilation and calculation based on data from Eurostat database. https://ec.europa.eu/eurostat/data/database, 10.01.2023.

areas in certain states, namely Belgium, Denmark, Germany, France, the Netherlands, Austria, and Slovenia, seem to manifest more pronounced social challenges compared to their rural regions.

A pivotal dimension distinguishing urban from rural areas, and influencing regional competitiveness variably, is human capital. Such disparities become particularly salient when focusing on quantifiable factors integral to competitiveness, such as education and digital skills. Among individuals aged 30–34, those possessing tertiary education constituted 50% in urban areas, dwindling to 33.5% in towns and suburbs, and further to a mere 28.4% in rural environments. Given the growing significance of digital skills in educational, entrepreneurial, and professional arenas, the lack of such skills can be detrimental.

Table 3.3 Unemployment rate in the EU-27 aged 15–74, in 2021 by degree of urbanization (in %)

EU member countries	Total	Cities	Towns and urban areas	Rural areas
EU-27 countries (from 2020)	7.1	7.9	7.0	6.0
Belgium	6.3	10.3	4.5	4.6
Bulgaria	5.3	3.5	5.3	8.9
Czechia	2.9	3.1	3.2	2.4
Denmark	5.1	6.8	4.9	3.7
Germany	3.6	4.7	3.3	2.3
Estonia	6.5	6.8	7.3	5.6
Ireland	6.3	6.8	7.0	5.6
Greece	14.9	13.8	15.9	15.2
Spain	14.9	14.5	15.9	13.7
France	7.9	8.9	8.5	6.3
Croatia	7.6	6.7	8.4	7.8
Italy	9.7	11.1	9.2	7.9
Cyprus	7.7	7.3	9.1	7.3
Latvia	7.9	7.8	7.5	8.1
Lithuania	7.4	6.0	8.6	8.6
Luxembourg	5.3	5.1	6.3	4.4
Hungary	4.1	3.0	4.0	5.3
Malta	3.4	3.5	3.3	:
The Netherlands	4.2	5.0	3.2	2.9
Austria	6.3	10.5	5.0	3.9
Poland	3.4	3.1	3.5	3.7
Portugal	6.7	7.3	6.1	6.5
Romania	5.6	2.8	5.8	8.5
Slovenia	4.8	5.8	4.3	4.7
Slovakia	6.9	4.2	6.6	8.6
Finland	7.8	8.9	7.7	6.1
Sweden	9.0	9.8	9.1	7.7

Source: Own compilation and calculation based on data from Eurostat database. https://ec.europa.eu/eurostat/data/database, 12.01.2023.

Regrettably, in 2019, only 48% of the population aged 16–74 in rural areas demonstrated basic or above-basic digital proficiency. In contrast, this proportion surged to 62% for urban inhabitants and remained at 55% for those residing in towns and suburbs. Notably, Belgium and Malta stood out as exceptions, recording the lowest digital proficiency within their cities. Such statistics underscore the tangible digital chasm between urban and rural areas within EU nations. The overall digital disparity across the EU spanned 14 percentage points, with this gulf widening to an alarming 23 percentage points in nations such as Bulgaria, Greece, Croatia, and Portugal in 2019 (Eurostat, 2020). This discourse accentuates the enduring territorial inequities, suggesting that economic and digital transformations, if harnessed appropriately, hold the potential to attenuate these territorial disparities, especially in the context of the prevailing rural–urban dichotomy.

Table 3.4 Persons at risk of poverty or social exclusion by degree of urbanization in 2020 (in %)

EU member countries	Cities	Towns and suburbs	Rural areas
EU–27 countries (from 2020)	22.0	19.8	22.8
Belgium	27.0	15.0	17.3
Bulgaria	23.0	29.9	46.7
Czechia	10.6	13.3	11.7
Denmark	20.3	14.1	12.7
Germany	25.6	19.0	18.9
Estonia	21.8	19.4	27.1
Ireland	18.3	22.6	20.1
Greece	26.2	28.3	32.4
Spain	25.1	26.3	29.1
France	22.9	19.5	13.1
Croatia	19.8	21.0	28.0
Italy	27.3	22.7	27.8
Cyprus	18.4	23.0	28.7
Latvia	22.7	26.6	29.4
Lithuania	18.5	27.7	29.9
Luxembourg	17.2	23.0	20.0
Hungary	13.7	17.4	22.2
Malta	19.3	16.7	:
The Netherlands	19.2	12.0	11.8
Austria	25.8	16.7	11.3
Poland	11.3	14.2	24.2
Portugal	17.1	20.1	24.4
Romania	14.1	24.4	45.4
Slovenia	17.8	14.2	14.5
Slovakia	8.1	16.0	17.2
Finland	15.6	15.1	17.5
Sweden	16.3	17.6	19.3

Source: Own compilation and calculation based on data from Eurostat database. https://ec.europa.eu/eurostat/data/database, 10.01.2023.

In the contemporary digital landscape, the availability of high-capacity telecommunications networks is paramount. The viability of business activities is intrinsically linked to the extent to which regions are equipped with robust broadband connections, a prerequisite for harnessing the myriad benefits of the digital economy. Recent data revealed that within the EU, 92% of urban households, 85% of rural households, and 90% in towns and suburbs had broadband subscriptions. However, a deeper dive into broadband connection speeds reveals a more nuanced disparity.

The EU has delineated clear targets for broadband speeds: a minimum of 30 Mbps for its general citizenry and an aspirational benchmark of over 100 Mbps for households. Regrettably, a significant proportion of EU states have not met these benchmarks. This discrepancy manifests itself in clear spatial patterns. Urban centers across the EU uniformly report speeds exceeding the 30 Mbps mark. Yet, a discernible divide emerges when comparing urban to rural areas concerning these speeds. In urban settings, a predominant portion of the populace consistently accesses speeds beyond 30 Mbps, with a notable segment even surpassing 100 Mbps. Conversely, in a majority of rural contexts in the EU, connection speed frequently languishes below the 30 Mbps threshold, albeit with a few notable exceptions. This delineates a pronounced urban–rural chasm in terms of broadband connection capabilities.

The urban–rural dichotomy extends beyond tangible indicators, permeating more abstract domains such as social capital and trust. These elements are increasingly recognized as vital drivers underpinning growth in contemporary economies. Within the EU, according to data from the Flash Eurobarometer on Regions, this trust divide appears more accentuated than disparities in trust toward national governments as recorded in the 2018 Standard Eurobarometer. While examining attitudes, an urban–rural divergence becomes apparent. However, this is not consistently evident across all categorizations, whether they be by country, region, or degree of urbanization. This urban–rural schism becomes manifest when evaluating electoral results from the 2014 and 2019 European Parliament elections, especially when votes are analyzed in the context of parties that hold varying stances on the EU and immigration issues (Scipioni and Trintori, 2021).

3.4 Conceptualizing smart cities and their contexts

The literature on smart city development, underpinned by ICT-empowered smart services and subsequent development of smart service ecosystem, accentuates the pivotal role of urban and rural cohesion. While the nexus between these paradigms is increasingly recognized, there is a manifest need for a robust conceptual framework. Such a framework would elucidate the intricate relationships and intersections between these phenomena, offering a more coherent understanding of their synergies. A well-structured conceptual mapping would not only enrich academic discourse but also provide a foundation upon which future empirical and theoretical studies can be anchored. Furthermore, with the heightened emphasis in public policy on fostering cohesion – exemplified notably by the EU's initiatives – it becomes even more imperative to progress conceptual groundwork. This would not only aid policymakers in crafting informed strategies but also ensure that the convergence of urban and rural landscapes through smart technologies is both efficient and equitable.

A smart service ecosystem, stemming from smart city development, seamlessly blends technological, human, and infrastructural elements to provide automated and intelligent services to its users. In this conceptual framework, the smart service ecosystem, through its essential components, serves as a bridge connecting urban and rural landscapes. These areas progressively merge into an ICT-enriched virtual realm, aptly termed as "smart territories," as illustrated in Figure 3.1.

As indicated by the above conceptual framework, the smart service ecosystem comprises eight essential elements that allow it to support a vast realm of

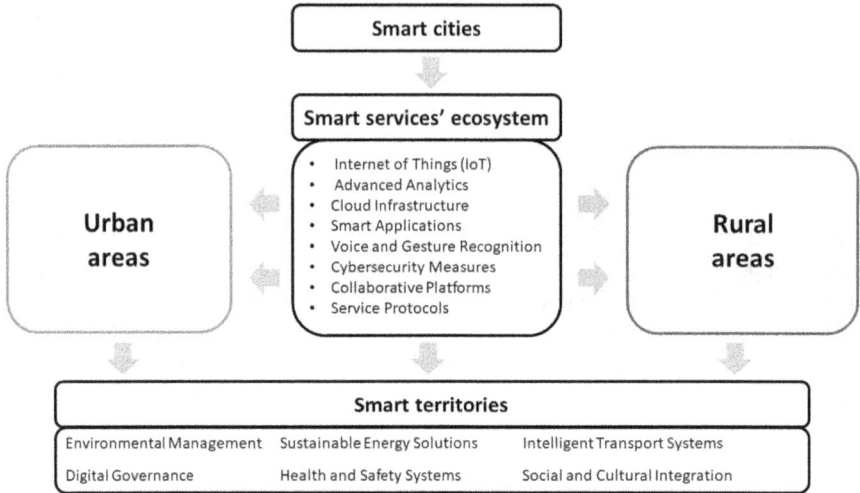

Figure 3.1 Conceptualizing smart cities and their contexts.
Source: Authors.

smart services. These components of smart service ecosystem include, as shown in Figure 3.1, interconnected devices, often termed the IoT, which collect, transmit, and receive data. Advanced analytics tools process this data, utilizing machine learning and artificial intelligence (AI) algorithms to derive insights and predictions. Cloud infrastructure supports the vast storage and computational needs, ensuring real-time responsiveness. Users interface with the ecosystem through smart applications, often enriched with voice or gesture recognition. Ensuring the security and privacy of user data is paramount; hence, robust cybersecurity measures are integral. Collaborative platforms facilitate interaction among stakeholders, while service protocols ensure the interoperability of different devices and platforms. Lastly, the ecosystem is continually evolving, driven by feedback loops, user needs, and technological advancements.

Smart territories, as an evolution of the smart city concept, encapsulate a more holistic integration of urban and rural environments leveraging advanced technology and data-driven approaches. While the exact components can vary based on the specific conceptual approach or research focus, in our conceptual framework, we have indicated six key components that provide characteristic features of smart territories as a merge of urban and rural areas. These components collectively work to transform traditional territories into interconnected, responsive, and sustainable environments for their inhabitants.

Smart territories aim to transform traditional landscapes divided along urban–rural split into areas that are interconnected, responsive, and tailored to provide sustainable environments for their inhabitants. At the heart of this transformation is environmental management, where systems are employed to diligently monitor and manage natural resources, waste, and pollution, ensuring both sustainable development and conservation. Complementing this is the emphasis on sustainable energy solutions that champion renewable energy sources, efficient storage, and smart grids. Intelligent transportation systems also see a revolution with the advent of connected transportation systems that seamlessly blend urban and rural travel through innovations of smart roads and autonomous vehicles. In this view, digital governance emerges as a cornerstone of smart territories, providing platforms for efficient public service delivery, fostering citizen engagement, and enabling real-time decision-making. Equally pivotal are the integrated health and safety systems, designed to cater to both bustling urban hubs and secluded rural areas. Lastly, the essence of these territories is enriched by initiatives that promote social and cultural integration, ensuring that technological advancements bolster, rather than overshadow, the unique local identities and values.

3.5 Discussion and conclusion

The evolution of smart cities has been significantly propelled by the incorporation of ICT-based solutions, which fueled the development of smart services, which have in turn facilitated the emergence of smart service ecosystems. These ecosystems can be envisaged as pivotal bridges, fostering a harmonious integration between urban and rural areas. To grasp the underlying mechanics of this urban–rural cohesion, it's imperative to delve into the intrinsic components of the smart service ecosystem. By analyzing these elements, one can discern the collaborative mechanisms that drive the convergence of diverse landscapes. Consequently, the culmination of this symbiotic union between urban and rural areas, facilitated by the smart service ecosystem, heralds the rise of "smart territories." These territories epitomize the seamless merger of technological advancement with varied geographies, crafting a future where spatial distinctions are bridged through innovation.

Addressing the research question of whether smart cities can serve as a bridge for the urban–rural divide, it becomes evident that smart cities play an instrumental role in fostering the development of the smart service ecosystem. This ecosystem, in turn, acts as a crucial link, amalgamating the urban and rural sectors. The conceptual framework delineates the key components of this smart service ecosystem that are pivotal in bridging this longstanding divide. These components encompass the following: IoT devices, advanced analytics, cloud infrastructure, smart applications, voice and gesture recognition, cybersecurity measures, collaborative platforms, and service protocols. IoT devices facilitate real-time data collection and transmission, advanced analytics provide data-driven insights, and cloud infrastructure ensures seamless data storage and processing. Additionally, smart applications offer user-friendly interfaces, while voice and gesture recognition technologies enable intuitive user interactions. Cybersecurity measures ensure the protection and integrity of the system, and collaborative platforms promote stakeholder interaction.

Lastly, service protocols guarantee the interoperability of devices and platforms, further enhancing the cohesive potential of smart cities.

Throughout the discussion in the chapter, the primary dimensions of disparities between urban and rural regions in the EU's cohesion strategy were examined. The EU's cohesion strategy is confronted by significant disparities between its urban and rural regions. It is exacerbated by diverse socioeconomic and digital dynamics. Inherent to the development of regions is the emergence of both challenges and opportunities from the digital transformation, more pronounced in urban settings. The results of our policy review indicate that urban regions consistently outperform their rural counterparts in GDP, net income, and broadband access, with cities often exceeding rural areas by significant margins. However, while cities may have higher GDP and net income, they also face higher unemployment rates and risks of poverty compared to rural areas. The human capital divide is evident too, with urban regions boasting a higher percentage of individuals with tertiary education and digital proficiency. Broadband speed disparities further highlight the digital divide, where urban centers predominantly achieve speeds beyond 30 Mbps, while many rural areas fall short. This dichotomy also extends into intangible areas like social capital and trust, with varying levels of trust and attitudes toward the EU evident across different regions. In sum, the EU's cohesion strategy faces an intricate challenge in bridging these urban–rural disparities, especially in the face of evolving digital landscapes.

When dwelling on the third research question, which probes the extent and way ICT-enhanced tools and applications influence a city's connection to its expansive geographic milieu, it is necessary to examine the pivotal components of the smart service ecosystem, as delineated earlier. These elements substantially shape urban–rural interactions. Contrary to a simplistic view that suggests a unilateral influence, wherein only rural areas are subject to transformation, the dynamics are more nuanced and multifaceted. As delineated in our conceptual framework, these smart territories possess distinctive features that have the potential to enhance the quality of life for their residents. Central to these enhancements are components that directly uplift inhabitants' quality of life. These encompass environmental management, sustainable energy solutions, intelligent transport systems, digital governance, health and safety systems, and social and cultural integration. Collectively, these elements reconfigure the urban–rural dynamic, with ICT playing a pivotal role in this transformation.

Future research avenues of the conceptual research undertaken in this chapter present a compelling tapestry of in-depth theoretical examinations and empirical pursuits. First, there is a pertinent need to apply the conceptual framework developed in this chapter, detailing the intricate interaction between smart cities and their ecosystems in the context of urban–rural divide, to specific case studies of regions and policies. This is especially salient in the context of the EU, where the dynamics of urban–rural divide are inevitably shaped by its cohesion policies. Second, a careful analysis of the smart service ecosystem components suggests that a fertile ground exists for research seeking to explore more granular mechanisms that bridge urban and rural divides. Here, the application of stakeholder analysis might provide illuminating insights. Lastly, the argument in the paper sought to conceptualize certain elements of smart territories, to highlight the implications of the confluence of urban and rural interactions. Thus, an exploration into how these outcomes could influence the quality of life of residents offers a promising direction. Such inquiries will not only enrich the academic discourse but also provide pragmatic insights for policymaking.

Acknowledgments

Research presented in this chapter is related to research conducted in the framework of the National Science Centre (NCN) grant "Smart Cities: Modelling, Indexing and Querying Smart City Competitiveness" (Nr DEC-2020/39/B/HS4/00579), led by Anna Visvizi.

References

Chakravarty, D., Goerzen, A., Musteen, M. and Ahsan, M. (2021) 'Global cities: A multi-disciplinary review and research agenda'. *Journal of World Business, 56*(3), p. 101182. doi: 10.1016/J.JWB.2020.101182

Cristofoletti, E., Gabriele, R. and Giua, M. (2023) 'Gaining in impacts by leveraging the policy mix: Evidence from the European Cohesion Policy in more developed regions'. *Journal of Regional Science*, doi: 10.1111/JORS.12666

Di Caro, P. and Fratesi, U. (2022) 'One policy, different effects: Estimating the region-specific impacts of EU cohesion policy'. *Journal of Regional Science, 62*(1), pp. 307–330. doi: 10.1111/JORS.12566

Döring, T. and Schnellenbach, J. (2006) 'What do we know about geographical knowledge spillovers and regional growth?: A survey of the literature'. *Regional Studies, 40*(3), pp. 375–395. doi: 10.1080/00343400600632739

Eurostat (2020) 'Urban and rural living in the EU'. https://ec.europa.eu/eurostat/web/products-eurostat-news/-/edn-20200207-1, 10.01.23.

Garcia-Ayllon, S. and Miralles, J. L. (2015) 'New strategies to improve governance in territorial management: Evolving from "smart cities" to "smart territories"'. *Procedia Engineering, 118*, pp. 3–11. doi: 10.1016/j.proeng.2015.08.396

Gong, D. and Shan, X. (2023) 'How does smart city construction affect urban–rural collaborative development? A quasi-natural experiment from Chinese cities'. *Land, 12*(8), p. 1571. doi: 10.3390/LAND12081571

Gutierrez-Velez, V.H., Gilbert, M.R., Kinsey, D. and Behm, J.E. (2022) 'Beyond the "urban" and the "rural": conceptualizing a new generation of infrastructure systems to enable rural–urban sustainability'. *Current Opinion in Environmental Sustainability, 56*, p. 101177. doi: 10.1016/J.COSUST.2022.101177

Haase, D., Larondelle, N., Andersson, E., Artmann, M., Borgström, S., Breuste, J., Gomez-Baggethun, E., Gren, Å., Hamstead, Z., Hansen, R. and Kabisch, N. (2014) 'A quantitative review of urban ecosystem service assessments: Concepts, models, and implementation'. *Ambio, 43*(4), pp. 413–433. doi: 10.1007/s13280-014-0504-0

Kashef, M., Visvizi, A. and Troisi, O. (2021) 'Smart city as a smart service system: Human-computer interaction and smart city surveillance systems'. *Computers in Human Behavior, 124*, pp. 106923. doi: 10.1016/J.CHB.2021.106923

Kędra, A., Maleszyk, P., Visvizi, A. (2023) Engaging citizens in land use policy in the smart city context'. *Land Use Policy*, *129*, p. 106649. doi: 10.1016/j.landusepol.2023.106649

Li, Y., Westlund, H. and Liu, Y. (2019) 'Why some rural areas decline while some others not: An overview of rural evolution in the world'. *Journal of Rural Studies*, *68*, pp. 135–143. doi: 10.1016/J.JRURSTUD.2019.03.003

Lillemets, J., Fertő, I. and Viira, A. H. (2022) 'The socioeconomic impacts of the CAP: Systematic literature review'. *Land Use Policy*, *114*, pp. 105968. doi: 10.1016/J.LANDUSEPOL.2021.105968

Luo, H., Zuo, M. and Wang, J. (2022) 'Promise and reality: using ICTs to bridge China's rural–urban divide in education'. *Educational Technology Research and Development*, *70*(3), pp. 1125–1147. doi: 10.1007/S11423-022-10118-8/TABLES/3

Lytras, M. D. and Visvizi, A. (2018) 'Who uses smart city services and what to make of it: Toward interdisciplinary smart cities research'. *Sustainability*, *10*(6), pp. 1–16. doi: 10.3390/su10061998

Malik, R., Visvizi, A., Troisi, O. and Grimaldi, M. (2022) 'Smart services in smart cities: Insights from science mapping analysis'. *Sustainability*, *14*(11), p. 6506. doi: 10.3390/SU14116506

Medeiros, E., Zaucha, J. and Ciołek, D. (2023) 'Measuring territorial cohesion trends in Europe: A correlation with EU Cohesion Policy'. *European Planning Studies*, *31*(9), pp. 1868–1884. doi: 10.1080/09654313.2022.2143713

Mettler, S. and Brown, T. (2022) 'The Growing Rural-Urban Political Divide and Democratic Vulnerability'. *Annals of the American Academy of Political and Social Science*, *699*(1), pp. 130–142. doi: 10.1177/00027162211070061/ASSET/IMAGES/LARGE/10.1177_00027162211070061-FIG3.JPEG

Navío-Marco, J., Rodrigo-Moya, B. and Gerli, P. (2020) 'The rising importance of the "Smart territory" concept: definition and implications'. *Land Use Policy*, *99*, p. 105003. doi: 10.1016/J.LANDUSEPOL.2020.105003

Palumbo, R., Manesh, M.F., Pellegrini, M.M., Caputo, A. and Flamini, G. (2021) 'Organizing a sustainable smart urban ecosystem: Perspectives and insights from a bibliometric analysis and literature review'. *Journal of Cleaner Production*, *297*, p. 126622. doi: 10.1016/J.JCLEPRO.2021.126622

Porru, S. et al. (2020) 'Smart mobility and public transport: Opportunities and challenges in rural and urban areas'. *Journal of Traffic and Transportation Engineering*, *7*(1), pp. 88–97. doi: 10.1016/J.JTTE.2019.10.002

Qiang, Q. and Jian, C. (2020) 'Natural resource endowment, institutional quality and China's regional economic growth'. *Resources Policy*, *66*, p. 101644. doi: 10.1016/J.RESOURPOL.2020.101644

Sano, Y., Hillier, C., Haan, M. and Zarifa, D. (2020) 'Youth Migration in the Context Of Rural Brain Drain: Longitudinal Evidence From Canada'. *Journal of Rural and Community Development*, *15*(4). https://journals.brandonu.ca/jrcd/article/view/1850 (Accessed: 8 October 2023).

Sawyer, L., Schmid, C., Streule, M. and Kallenberger, P. (2021) 'Bypass urbanism: Re-ordering center-periphery relations in Kolkata, Lagos and Mexico City'. *Environment and Planning*, *53*(4), pp. 675–703. doi: 10.1177/0308518X20983818/ASSET/IMAGES/LARGE/10.1177_0308518X20983818-FIG3.JPEG

Scipioni, M. and Trintori, G. (2021) 'A rural-urban divide in Europe? An analysis of political attitudes and behaviour'. *JRC technical report*, Office for Official Publications of the European Union 2021. https://publications.jrc.ec.europa.eu/repository/handle/JRC123124 (Accessed: January 13, 2023)

Troisi, O., Grimaldi, M. and Monda, A. (2019) 'Managing smart service ecosystems through technology: How icts enable value cocreation'. *Tourism Analysis*, *24*(3), pp. 377–393. doi: 10.3727/108354219X15511865533103

Urbano, D., Aparicio, S. and Audretsch, D. (2019) 'Twenty-five years of research on institutions, entrepreneurship, and economic growth: what has been learned?'. *Small Business Economics*, *53*(1), pp. 21–49. doi: 10.1007/S11187-018-0038-0/TABLES/6

Vergara, L. and Salazar, G. (2021) 'Non-metropolitan cities in Latin American urban studies: Between trickle-down urban theory and singularisation theory'. *International Development Planning Review*, *43*(3), pp. 321–344. doi: 10.3828/idpr.2020.18

Visvizi, A. and Lytras, M. D. (2018) 'Rescaling and refocusing smart cities research: from mega cities to smart villages'. *Journal of Science and Technology Policy Management*, *9*(2), pp. 134–145. doi: 10.1108/JSTPM-02-2018-0020/FULL/PDF

Visvizi, A., Mazzucelli, C.G., Lytras, M. (2017) 'Irregular migratory flows: Towards an ICTs' enabled integrated framework for resilient urban systems'. *Journal of Science and Technology Policy Management*, 8(2), pp. 227–242. doi: 10.1108/JSTPM-05-2017-0020

Wirtz, B. W. and Müller, W. M. (2023) 'An integrative collaborative ecosystem for smart cities — A framework for organizational governance'. *International Journal of Public Administration*, *46*(7), pp. 499–518. doi: 10.1080/01900692.2021.2001014

Zöllner, S., Heidinger, M., Sager, S., Lüthi, S. and Thierstein, A. et al. (2023) 'Advanced producer services in Germany: A relational perspective on spatial core–periphery structures, 2009–19'. *Regional Studies, Regional Science*, *10*(1), pp. 347–368. doi: 10.1080/21681376.2023.2185536

4 Smartification, quality of life, and the challenges of urbanism

The case of the Line City

Abeer S. Y. Mohamed

4.1 Introduction

Cities play a key role in transitioning to a more sustainable future that interacts positively with climate change. Smart cities (Visvizi & Lytras, 2018) have served as a central concept in interactive, responsive urban management for the 21st century using information and communication technology (ICT) (Albino et al., 2015). Urban smartness is based on the idea that cities can become more sustainable and eco-efficient while promoting the well-being of citizens (Visvizi & Pérez del Hoyo, 2021). Smartification refers to applying information technology in various stages of planning, designing, building, and managing cities (Guido et al., 2022). In parallel with these considerations, debates are emerging about post-humanism and non-human-centered design in the context of smart city emergence (Luusua et al., 2017). Smart city studies refer to using technology to enhance the efficiency and effectiveness of various city services (Visvizi & Pérez del Hoyo, 2021). In line with this concept, it is very important to involve smart computing infrastructures and technologies to improve the services provided by multiple city agencies (Yin et al., 2015). Smart cities use technology and human and machine resources to improve their services (Visvizi, 2023). Another important pillar that should be included in the conversation on smart cities, quality of life, and urban planning is that of a human-centered design. The latter provides all requirements for society, respects its needs, and interacts with it effectively. Human-centered design in the smart city context may be applied through an interactive, empathic city approach, which places the human being and, thus, an individual's personal experience in the center of the urban planning process.

Urban development today attests to the mechanism of the twin processes of digitalization and smartification unfolding in the context of the smart city agenda. This main context reflected on the urban design paradigm starting from the governmental bodies, codes, applications, societies, and all related fields and how these processes relate to the concepts of smart cities. This chapter examines the important relationship between smartification and the interactive, empathetic quality of life (IEQoL) by focusing on the city, transitioning to a smart city; quality of life, viewed as the ultimate objective of urban planning; and smartification, viewed as a complex process associated with the evermore pervasive use of digital solutions in our lives, including the city. To

DOI: 10.1201/9781003415930-6

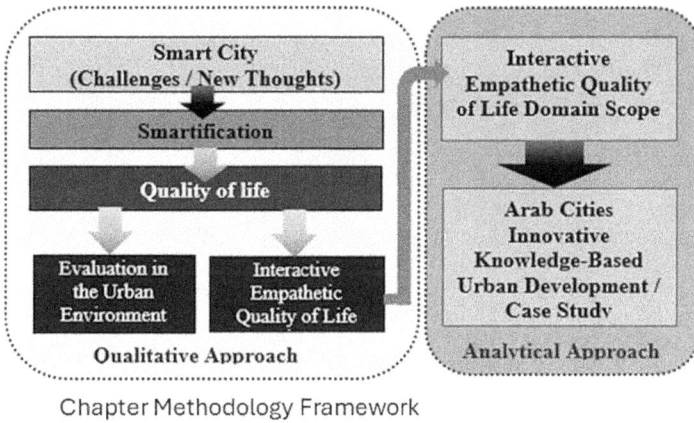

Chapter Methodology Framework

Figure 4.1 Chapter methodology framework.

identify and understand the mechanisms that govern the relationship between these three sets of factors, i.e., the quality of life, urban planning, and smartification, a research model (Figure 4.1) is devised. It is then applied to the examination of the Line City. Discussion and conclusions follow.

4.2   The smart city dilemma

A smart city is a developed urban region that excels in a variety of critical areas such as economy, mobility, the environment, people, and governance. Global cities are clamoring to become "smart" by increasing their residents' well-being and quality of life, the environment, people, and governance. Excellence in these critical areas can be attained by integrating technological innovation with the most modern ICT infrastructure (Visvizi & Lytras, 2018). Given the variety of smart city models and ways of applying them across the world, a smart city dilemma appeared consistent with the question of how to define a smart city, given the immense variability of circumstances and conditions.

A smart city is designed to meet urban challenges and provide residents and visitors with the following amenities (Ibrahim et al., 2018; Gurstein & Hutton, 2019). Rapid urbanization is a major issue in 2050, with two-thirds of the world's population living in cities and 25% of all city dwellers living in one of 70 megacities (Gurstein & Hutton, 2019). To address this, transportation systems need to be updated with different types and techniques (Suma et al., 2018), integrated urban mobility needs to be managed, infrastructure needs to be upgraded, safety and security need to be addressed, and green consciousness needs to be addressed (Yang et al., 2019). These changes are necessary to ensure the safety and security of roads and public institutions, environmental issues, climate change, and energy efficiency (Ibrahim et al., 2018).

Smart cities offer promising solutions using digital technologies and services (Figure 4.2). Digital technologies and services improve residents' lives by

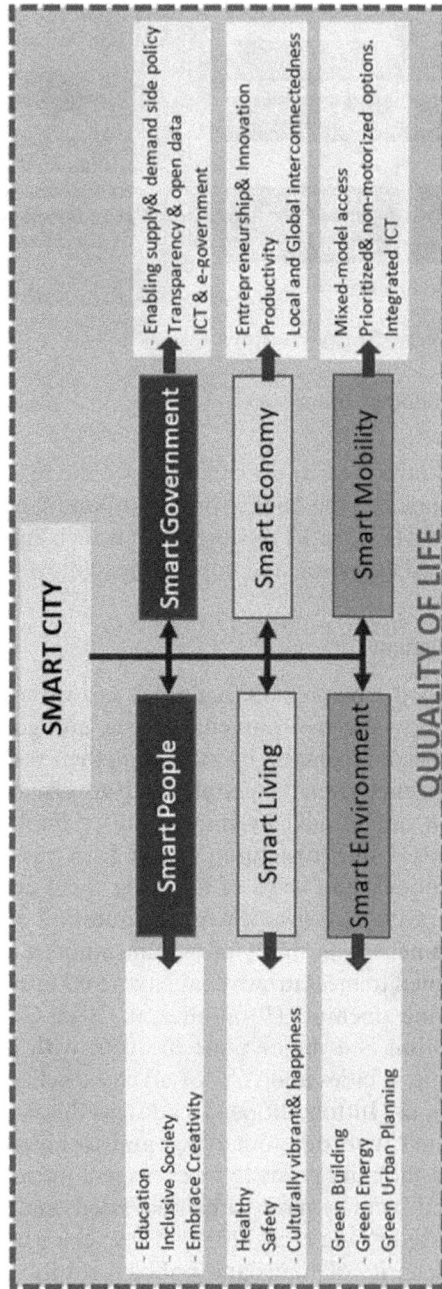

Figure 4.2 Smart city features, arranged by the authors based on insights from Cohen (2012) and Kumar & Dahiya (2017).

providing new experiences of city services through enhanced digital use cases (Gade, 2019; David et al., 2022). Mining city data will enable city entities to make informed decisions and improve city management (Caragliu & Del Bo, 2019). A "smart" environment positively affects the city economy through direct and indirect contributions. Six characteristics are the cornerstone for further developing smart cities: education, cosmopolitanism, open-mindedness, involvement in civic affairs, skill and learning development.

Smart economies contain productivity, entrepreneurship, innovation, human capital quality, labor market, and global. The smart environment includes air quality, sustainable management of resources, energy efficiency, and attractiveness to nature (Cugurullo, 2018). Smart living includes housing quality, educational resources, individual safety, social cohesion, cultural amenities, and health conditions (Suma et al., 2018). Smart mobility includes infrastructure and connectivity, accessibility across borders, and system sustainability (Siano, et al., 2018) (Kamel, 2013). Smart governance includes access to public services, participation in decision-making, efficient and transparent administration, and innovation in service provision (Yin et al., 2020).

4.3   Smartification

Smartification is about building new technologies and finding ways to include residents and help them achieve a better quality of life (Siano et al., 2018). This requires a shift in mindset from focusing on technology to the people who will use it daily. Urban smartification refers to a new concept that uses technology to help improve the quality of life and urban development. According to the technical perspective of David et al. (2022), they identified seven major strategies necessary to develop a smart city concept, which is as follows (Guido et al., 2022): entity framework with context awareness for environmental intelligence, acceptable user interface, relationship to regulations, object-to-object communication, location-based services (LBSs) with numerous exchange settings, automatic adaptability to new work or personal situations, and Enhanced cybersecurity measures. These features allow for object-to-object communication without human intervention, exchange of settings based on environmental context, and automatic adaptability to new work or personal situations.

Aligning with the main aim of this study about the twin processes of digitalization and smartification, smartification is still being developed, so there are many ways to define it. A smart city is where all ICT possibilities can be used, including data accessibility, data processing, and information exchange in static and mobile situations. This should be reflected in all processes, planning, development, and daily life in the governmental and private sectors. A smartification scale measures a city's quality of life (Deguchi, 2020). It was designed to help urban planners and architects select the best strategy for smartification, which can then be implemented in each city according to its needs. The smartification scale has four pillars (Caragliu & Del Bo, 2019): smartification level and components, infrastructure, social and cultural aspects, and environmental aspects.

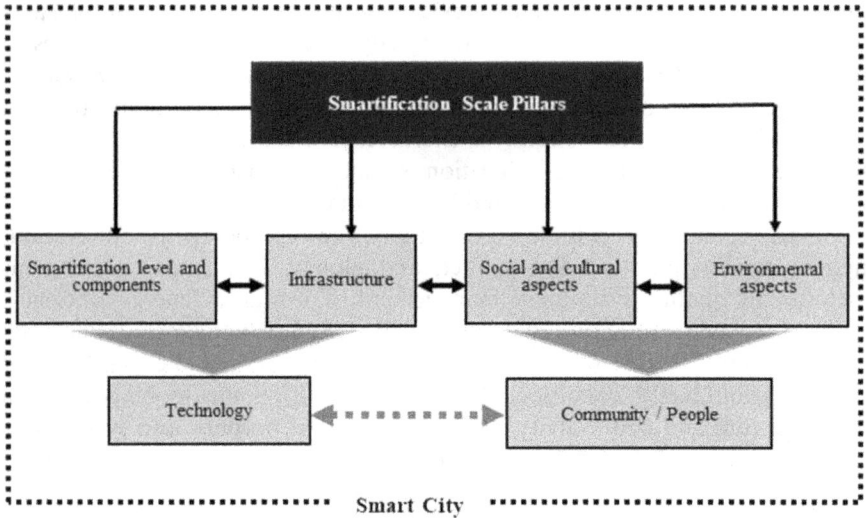

Figure 4.3 Smartification scale has four pillars.

These pillars (Figure 4.3) must be considered when designing a city's space. Each pillar has several indicators with various levels of importance that must be analyzed before taking steps toward smartifying any part of the community or city.

The first step to smartifying any part of a community is to measure its current level of smartification. This will allow for determining what needs to be improved and where efforts should be focused. Not only that, but the most important is society's ability to accept and use this smartification and to be more familiar with using it and any future adaptation or upgrading.

The sustainable development goals (SDGs) and smart cities are two influential policymaking concepts that influence discussions today, making cities inclusive, secure, resilient, and sustainable. The four-pronged aims of inclusion, safety, resilience, and sustainability are sometimes difficult for local authorities to execute since they frequently lack the authority, funds, and expertise to do so (Visvizi & Pérez del Hoyo, 2021). Self-organization is based on interactions between parts and feedback mechanisms in natural systems. Rather, smart citizens evaluate, recommend, interpret, and share experiences, give feedback, and participate actively in creating smart urbanism. This new urban awareness transforms humans into remote sensor-generating cybernetic data points and information flows (Allam & Newman, 2018). It may be an everyday reality in our cities that intelligently controlled lighting systems are designed based on residents' experiences in urban spaces.

The British Standards Institute (BSI) and the British Business, Innovation, and Skills (BIS) Department have developed a transition framework to facilitate the transition to smart, sustainable cities (SSCs) in the United Kingdom. The proposed framework (Figure 4.3) consists of four top-level components named (BSI, 2014): the planning, application, delivery, consolidation, and conversion phases of a city-specific roadmap are necessary to ensure the business case is fully presented and all key stakeholders are involved. The planning phase includes the guiding principles, city vision, benefits realization framework, and the SSC roadmap. The application stage focuses on maximizing benefits and minimizing delivery risks by focusing on quick wins. The delivery period includes the first wave of smart services and applications. The consolidation phase focuses on long-term strategic solutions to drive changes toward adopting original SSC applications and services and learning from intelligent data and user feedback. The conversion phase includes building on a broader range of SSC projects and completing the transition to a full strategic IT platform.

The European Platform for Intelligent Cities (EPIC, 2013) proposes an SSC transformation framework/roadmap with manual deployment stages diagnosed for the EPIC framework to assist towns in broadening their SSC vision. These stages include outlining concrete undertaking plans, putting force SSC solutions, and functioning SSC offerings in a cloud environment (Figure 4.4). The visioning, planning, design, construction, and delivery phases are essential for successful SSC services development. These phases involve gathering business requirements, deploying and testing fixed asset services, and preparing for business transformation.

4.4 Quality of life evaluation in the urban environment

The quality of life is made up of three basic regions (Rosenbaum et al. 2007) and environments (Cugurullo, 2018): life is composed of physical, mental, and intellectual components, while property rights connect people to social, physical, and network environments, and group discussions and engagement are essential for achieving goals. Quality of life is associated with individuals' happiness' levels and enjoyment of numerous factors of existence, including occupation, housing, network, existence, and society, as well as economic, family, spiritual, recreational, social, emotional, and spiritual well-being. Happiness no longer best reflects the existence of individuals but also people's beliefs about their life in the area (Keles, 2012). The diversity and demographics of individuals assessing fundamental needs and expectations, along with their attitudes and satisfaction levels towards services available for their enjoyment, are crucial factors in evaluating the quality of life. This consideration is a fundamental principle in urban planning and integral to sustainable development. It can meet the world's present system's needs without neglecting future generations' responsiveness (Serag El Din, 2013). Satisfying individual needs (Figure 4.4) not only is a prerequisite for sustainability but also creates human happiness, which is one important goal in achieving a high quality of life (Uzzell & Moser, 2006).

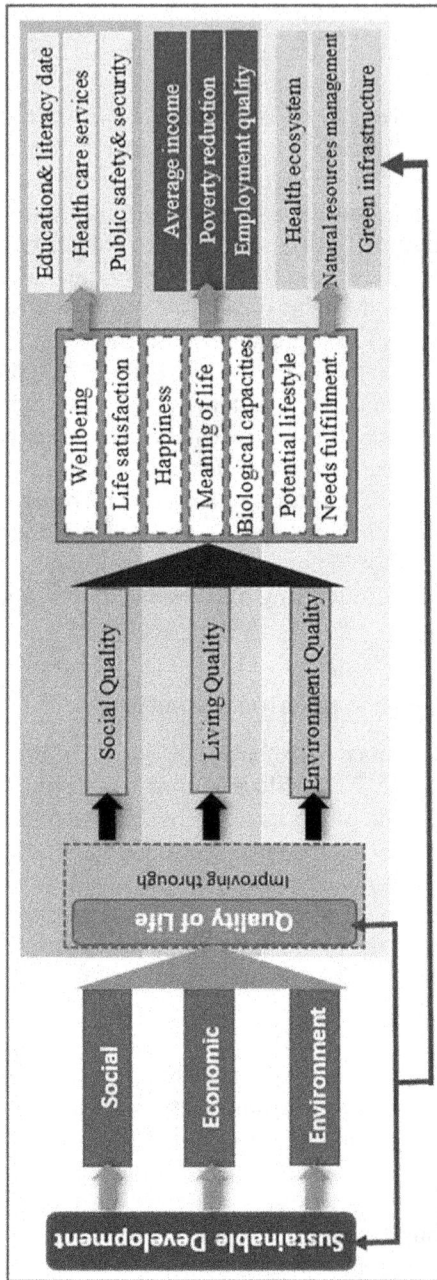

Figure 4.4 Sustainable development and the concept of QOL are conceptually similar.

Source: Author.

4.4.1 *Interactive empathetic quality of life: From smart to empathetic city*

By analyzing smart cities and their response to residency needs with different aspects and aligning with quality of life, this study presents the new quality of life urbanism proposal: IEQoL city – means a city with human–environment–social interaction as the core. This can be achieved by understanding the evolution of smartifications and IEQoL city through the conceptual framework of smartification and quality of life in pursuing urbanism challenges that this study presents.

This study proposes a systematic-thinking approach focusing on the integrated clear vision of smartification systems rather than parts. It argues that each sector of smartness must reexamine how to work together rather than being interested in self-glorifying data profiles of their own operations and impacts; thus, smartification becomes the new iconic.

4.4.1.1 *Humanitarian value as centralized core*

The existence's comfort is made up of the following: existence refers to physical, mental, and intellectual components, while property rights connect people to social, physical, and network environments and have become a focus for achieving goals and recreations (Kamel, 2013; Brenner, et al., 2012). The quality of life is associated with individuals' happiness levels and enjoyment of various factors of existence, including occupation, housing, network, existence, and society. Developing collaborative organizations and reducing the price of associated basic common goods are examples of possible impacts of harnessing an empathic future (Keles, 2012; Brenner et al., 2012).

4.4.1.2 *Interactive empathetic quality of life domain scope*

The IEQoL city domain scope has ethical grounds for social justice and equity (Gospodini & Manika, 2020; Sorrell, 2007). Holistic well-being via healthy collaboration within moral grounds is critical for outlining a common cause, culture, and value systems that constitute an IEQoL city (Gomes et al., 2018). Improvements in energy efficiency often encourage greater use of the services that energy helps provide. Citizen initiatives such as education opportunities, access to ICT, cultural activities, economic development, infrastructure provision, participatory budgeting, and community planning can be beneficial in creating urban commons (Cugurullo, 2018). The digital literacy of citizens is a key factor to consider when developing technology-intensive strategies to understand public needs and demands. Promoting alternative data ownership regimes will be the needed target (Prado et al., 2016).

4.5 Smart cities quality of life, collective smartification, and place identification (Figure 4.5)

Cities adopting "smartness" may solve increased energy consumption, traffic congestion, and social inequalities. Technology should not only eliminate the need to access and engage in the fundamental activities required to support an urban existence but also contribute to ensuring a good quality of life by employing policies (Bawany et al., 2015; Gospodini & Manika, 2020). The "15-Minute City" idea argues for higher closeness and social contact captured by the "density" dimension, digitization, and diversity pillars, resulting in more tightly woven community fabrics (Khan & Bele, 2016). This would be aided by growing technical breakthroughs, which have resulted in the birth of creative urban planning models such as the smart city idea (Li et al., 2021). The proposed planning approach stresses the four components and provides a viable alternative. Allowing individuals to transfer quickly from residential, employment, and commercial sectors to educational institutions (Adkins et al., 2019; Serag El Din, 2013), health facilities, and other fundamental resources must be linked to different dimensions to create the desired urban change.

The city idea has two components:

- Mixed-use districts need a good balance of residential, business, and entertainment elements (Bawany et al., 2015). Purchasing mixed-use areas are critical to achieving optimal density and closeness to important services in the 15-minute city model.

For example, under the smart city idea, numerous platforms, including digital ones (McKinsey & Company, 2018), such as Moreno's suggested concept, support elements such as inclusion, participation, and the provision of real-time services (Dembski et al., 2020) (Lea, 2017). Incorporating these elements into the smart city idea promotes the effective use of numerous technologies, with comparable far-reaching ramifications for the future.

The proposed integrated smart city interactive shareable framework methodology in this chapter (Figure 4.5) can meet the fundamental application needs and quick creation of theme systems required by government agencies, corporations, and institutions (Kamel et al., 2015) and the public. Unlike universal application systems, the system architecture described here is tailored to specific users, with various application modes available, such as direct application, custom application, standard services, and inline calls. A circular loop represents the four-dimensional (4D) smart city and cybernetic management system. Services, infrastructure, locations, and elements of the lifeworld are all interconnected in smart cities. The main building components of the system are urban governance (the regulator) and city-scale practices (the program and plans).

- *Direct application*: enables regular users to use various service functions, import theme data, and perform geographical statistical analyses and queries.
- *Customized program*: dedicated to serving custom users who combine a secondary development interface with their needs to encapsulate a bespoke graphical interface. It also enables the development and extension of functionality to tailor theme application systems.

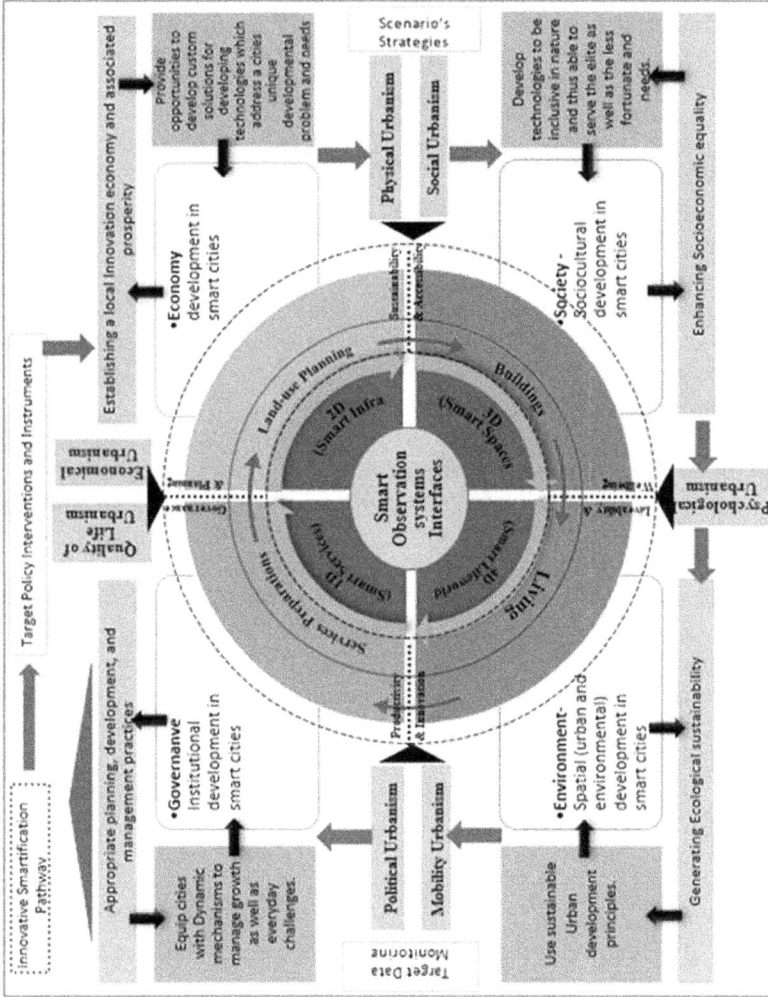

Figure 4.5 Smart city: a shareable framework design methodology for "Smartification".

Source: Author.

- *Conventional customer support*: primarily serving development users by assisting them in the implementation of map and geographic information network services (Bawany et al., 2015), helping the client to use other Geographic Information System (GIS) software to develop thematic application systems, and performing the distributed call of service resources (Lea, 2017).
- *Inline call*: primarily supporting embedded users by offering network connecting technology to assist users in completing an online call of other information services without requiring changes to the company operating systems currently in use.

4.6   Arab cities innovative and have knowledge-based urban development

Cities in the Arab world are entrepreneurial and are developed based on innovative research. In the Arab world, SSCs must overcome four obstacles: adult illiteracy and political engagement, data collection, validity, and reliability. Analytic procedures are essential for decision-making and bound rationality impacts the plan. Establishing an SSC in a country where most people are illiterate and impoverished is difficult. To become SSC, conventional towns and cities must undergo new developments. Transferring technology that has already been developed elsewhere will not be successful. A successful shift to SSC depends on the environment and how creative traits like organizational enablers, culture, and leadership are present. From the linear layout of individual structures to regional band development to the architectural scale to regional planning, linear cities vary in size. They are moving from focusing on a single public transportation line to more intricate multiline strip systems (Batty, 2022). It is vital to introduce systematics in linear geometry and applications of the linearity concept in urban planning because the differentiation of scales at which the idea of linearity and banding is implemented may lead to confusion. The following traits can be used to define the size of linear urban developments (Tufek & Stachura, 2019): comparable linear urban forms, linear communities, linear cities, linear megastructures, transit-oriented developments, and infrastructure corridors.

The concept of a linear city as a geometrically shaped metropolitan area has appeared in many forms and variations throughout the history of urban design. Planning linear or strip architectural and urban layouts is a significant category of situations (projects). Such "linear" city systems are ones where "the ribbon of buildings is narrow enough to limit pedestrian traffic in the transverse direction" in terms of "linear" cities. The ideal city form is one in which it takes the fewest steps to get from one house to all the others. The linear city development's reasons, causes, and goals influence yet another typological divide – between urban and rural areas (Batty, 2022), equal access to open areas and bodies of water, a concentration of buildings between cities, convenient access to the transit road, optimum utilization of public transportation, and no-collision zones for pedestrians and vehicles are essential.

The linear city is a compact development system focused on mass communication. The most prevalent challenges and issues facing metropolitan regions today that a linear city may address are air pollution, noise pollution, and overspill of urban landmass (urban sprawl) by some urban areas (Tufek & Stachura, 2019): traffic accidents, excessive traffic congestion, air pollution, municipal infrastructure failures, public transportation accessibility, access to basic and higher-order services, employment, and green spaces, and degenerate buildings all contribute to traffic accidents, congestion, air pollution, and municipal infrastructure failures.

Vertical urbanism involves viewing the cities from a unique perspective to break down the barriers between buildings and cities. Start considering what constitutes a great city; they may include wonderful locations to live, wonderful places to buy and dine, viable businesses, ideally close to where people live, culture and entertainment, parks, and open spaces (Abdelsalam et al., 2018), and well-connected public transportation. With more vertical master planning and horizontal connection, all the excellent aspects of the city's texture and fabric may be arranged. Urban structures that combine city life and vitality, such as vertical hybrids, can include housing, hotels, offices, cultural venues, retail, parks, and open spaces.

4.6.1   *The case of the Line City: The background and the content*

Based on best practices in SSC deployments, Key Success Factors (KSFs) for cities have been identified and advocate for an all-encompassing citywide strategy. The most important details in this text are a city's vision and long-term approach, strong leadership, accessible and shared infrastructure, innovative business structures, ecosystem characteristics, adoption processes, communication and advertising strategies, and community consciousness. A city's vision and long-term approach serve as a clear guide to direct all its activities toward a shared objective. Strong leadership is necessary to lead and coordinate several simultaneous activities and verticals without running the risk of setting up silos. Accessible and shared infrastructure is necessary to enable better service delivery and data monetization options. Innovative business structures are necessary to transition from traditional hardware/products to cutting-edge services and solutions, value chain disruption across industries, and technological advancements. Ecosystem characteristics include active citizen participation and third-party involvement. Adoption processes include an overall corporate scenario prioritization exercise to identify the most crucial activities, and community consciousness is essential to make people aware of the city's services and how residents may use them (Albalawi et al., 2019; Alfawzan et al., 2020). The Line City is a new, futuristic, emission-free city in Saudi Arabia in the form of a 170-kilometer line, announced by Mohammed bin Salman, Crown Prince and Chairman of the Neom Company Board of Directors (Alfawzan et al., 2020), aroused a discussion about ideal cities, the geometry of a linear city, and its legitimacy. So, it is essential to refer to urban history and look at similar ideas of linear cities and the attempts to implement them for over 150 years

(Neom, 2021). Although the linear city has not been fully realized according to the initial assumptions of the concept, today, it turns out that the impossible will become a reality.

The Line is a $500 billion cross-border city that will use innovative technology (Alkeaid, 2018). It will consist of a 170-km-long network of towns related to a high-speed subterranean train and/or a hyperloop. The uppermost layer, the Pedestrian Layer, will be exclusively for pedestrians. No journey is expected to take more than 20 minutes (PR Newswire, 2017; Karabell, 2017). The main features of sustainable urbanism are the IEQoL and smartification of the Line City is a revolution in urban living (https://www.neom.com/en-us/regions/theline): Vertical urbanism is a model for nature preservation, reducing infrastructure footprint, alternative ways to live, and high-speed linear mobility. Around 95% of land and sea are protected by nature.

They were reflected in the design of interactive responses for enhancing the quality of life and formulized The Line City to be an international city leading the world in future urbanism and sustainable smart cities as in Figure 4.6 (PR Newswire, 2017).

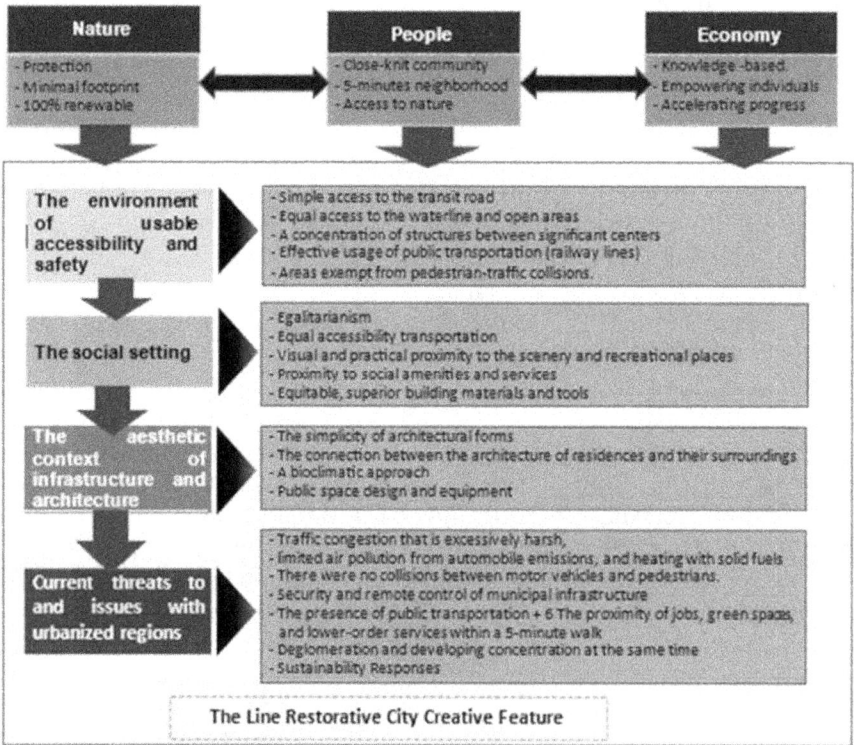

Figure 4.6 The Line City analysis key features.

Source: Author.

The most important details in this text are the adoption of artificial intelligence (AI) for life enhancement, the saving of 95% of Neom nature, the total renewable energy system, the longest journey time end to end, three separate layers for services and circulation, an invisible layer for infrastructure, new smart mobility via ultra-high-speed autonomous transportation, and the creative economy, which attracts investment and talent and creates 380,000 jobs in the future. These details demonstrate how AI can be used to improve quality of life, save 95% of Neom nature, and develop communities.

4.7   Conclusions

The Line City is a unique experience that will positively impact urban planning and thoughts and reorient them to new phases of high quality of life and the importance of making humans the main focal point of any development. This city emerged from an integration of the twin processes of digitalization and smartification through the urban design paradigm started from the governmental bodies, codes, applications, societies, and all related fields. Through the analysis of the Line city sustainable urbanism's main features, it is considered a unique case study for an IEQoL and smartification city and a revolution in urban living that will be motivating for future projects that will make real the Kingdom Saudia Arabia the leader of smartification and innovation for the entire world. Aligning with the ambitious aspirations framework of Vision 2030 for transforming the Kingdom of Saudi Arabia into a unique leading global model with different life features, especially with an emphasis on future city development.

The Line City offers a way of living that prioritizes the needs of people over all else. AI technology is significant and profoundly affecting in this context due to its best geographic position, distinguished by beautiful natural surroundings, renewable energy, and easily accessible services. According to the perspective of the Kingdom of Saudi Arabia, AI (smartification) is reshaping urban life and transforming the nation. This kind of existence is impossible to create using just hydrocarbon fuels. We may thus predict, because of this research, that the Kingdom of Saudi Arabia will see a qualitative jump in its technological growth in the future. Interactive, empathetic quality of life and smartification city is the new pass for the smart, sustainable city that should be a guide to follow in urban planning. Integrating smart city initiatives should be ingrained within the core of a nation's forward-thinking developmental strategy. Consequently, the implications of smartification and the emergence of intelligent entities, achieved through collaborative efforts and the seamless integration of public, private, and voluntary sectors, play a crucial role in shaping the spatial evolution of smart cities.

References

Abdelsalam, A.E., et al., (2018). Sustainable Vertical Urbanism as a Design Approach to Change the Future of Hyper Density Cities Redesigning the Skyscraper from the Urban Design Perspective. *Journal of Advance Research in Mechanical & Civil Engineering*, 5(7), Jul. https://doi.org/10.53555/nnmce.v5i7.300

Adkins, L., et al., (2019). Class in the 21st Century: Asset Inflation and the New Logic of Inequality. *Environment and Planning A: Economy and Space*, 51, 309–332.

Albalawi, H., et al., (2019). Energy Warehouse - A New Concept for NEOM Mega Project. *2019 IEEE Jordan International Joint Conference on Electrical Engineering and Information Technology (JEEIT)*. IEET. https://doi.org/10.1109/JEEIT.2019.8717480

Albino, V., et al., (2015). Smart Cities: Definitions, Dimensions, Performance, and Initiatives. *The Journal of Urban Technology*, 22, 3–21.

Alfawzan, F., et al., (2020). Wind Energy Assessment for NEOM City, Saudi Arabia. *Energy Science Engineering*, 8(3), 755–767. https://doi.org/10.1002/ese3.548

Alkeaid, M.M.G., (2018). Study of Neom City Renewable Energy Mix and Balance Problem [Unpublished Masters' thesis]. KTH Royal Institute of Technology School of Architecture and the Built Environment.

Allam Z., & Newman P., (2018). Redefining the Smart City: Culture, Metabolism and Governance. *MDPI*, 1, 4–25.

Batty, M., (2022). The Linear City: Illustrating the Logic of Spatial Equilibrium. *Computational Urban Science*, 2, 8. https://doi.org/10.1007/s43762-022-00036-z

Bawany, N.Z., et al., (2015). Smart City Architecture: Vision and Challenges. *International Journal of Advanced Computer Science and Applications*, 6, 246–255.

Brenner, N., et al., (2012). *Cities for People, Not for Profit: Critical Urban Theory and the Right to the City*. Routledge, New York.

BSI (2014). Smart Cities – Vocabulary. British Standards Institute (BSI), BSI Standards Publication, PAS 180:2014. London: United Kingdom.

Caragliu, A., & Del Bo, C. F., (2019). Smart Innovative Cities: The Impact of Smart City Policies on Urban Innovation. *Technological Forecasting and Social Change*, 142, 373–383. https://doi.org/10.1016/j.techfore.2018.07.022

Cohen, B. (2012). 6 Key Components for Smart Cities. *UBM's Future Cities*. http://www.ubmfuturecities.com/author.asp?section_id=219&doc_id=524053

Cugurullo, F., (2018). Exposing Smart Cities and Eco-Cities: Frankenstein Urbanism and the Sustainability Challenges of the Experimental City. *Environment and Planning A: Economy and Space*, 50(1), 73e92.

David, B., et al., (2022). Design Methodology for "Smartification" of Cities: Principles and Case Study. *2022 IEEE 25th International Conference on Computer Supported Cooperative Work in Design (CSCWD)*, May. https://doi.org/10.1109/CSCWD54268.2022.9776257

Deguchi, A. (2020). From Smart City to Society 5.0. In: *Society 5.0*. Springer, Singapore. https://doi.org/10.1007/978-981-15-2989-4_3

Dembski, F., et al., (2020). Urban Digital Twins for Smart Cities and Citizens: The Case Study of Herrenberg, Germany. *Sustainability*, 12, 2307.

EPIC. (2013). EPIC Roadmap for Smart Cities. European Union, European Platform for Intelligent Cities (EPIC), Version 1.0, Project no. 270895.

Gade, D., (2019). Technology Trends and Digital Solutions for Smart Cities Development. *International Journal of Advance and Innovative Research*, 6(1, Part – 4), 29–37.

Gomes, E., et al. (2018). Assessing the Effect of Spatial Proximity on Urban Growth. *Sustainability*, 10, 1308.

Gospodini, A., & Manika, S. (2020). Conceptualising "Smart" and "Green" Public Open Spaces; Investigating Redesign Patterns for Greek Cities. *Civil Engineering and Architecture*, 8(3), 371–378. https://doi.org/10.13189/cea.2020.080322

Guido, G., et al., (2022). Prioritizing the Potential Smartification Measures by Using an Integrated Decision Support System with Sustainable Development Goals (a Case Study in Southern Italy). *Safety*, 2022(8), 35. https://doi.org/10.3390/safety8020035

Gurstein, P., & Hutton, T., (2019). *Planning on the Edge: Vancouver and the Challenges of Reconciliation, Social Justice, and Sustainable Development*. UBC Press, Vancouver, BC. https://www.neom.com/en-us/regions/theline

Ibrahim, M., et al., (2018). Smart Sustainable Cities Roadmap: Readiness for Transformation Towards Urban Sustainability *Sustainable Cities and Society*, 37, 530–540. https://doi.org/10.1016/j.scs.2017.10.008

Kamel Boulos, M. N., et al., (2015). Social, Innovative and Smart Cities are Happy and Resilient: Insights from the Who Euro 2014 International Healthy Cities Conference. *International Journal of Health Geographics*, 14, 1–12.

Kamel, M., (2013). Encouraging Walkability in GCC Cities: Smart Urban Solutions. *Smart and Sustainable Built Environment*, 2(3), 288e310.

Karabell, Z. (2017). Saudi Prince Plans a "City of the Future". Don't Bet on It. *Wired*. https://www.wired.com/story/saudi-prince-plans-a-city-of-thefuture-dont-bet-on-it/

Keles, K., (2012). The Quality of Life and the Environment. *Procedia-Social and Behavioral Sciences*, 35, 23–32. https://doi.org/10.1016/j.sbspro.2012.02.059

Khan, S., & Bele, A. (2016). Transforming Lifestyles and Evolving Housing Patterns: A Comparative Case Study. *Open House International*, 41(2), 76–86. http://www.openhouseint.com/abdisplay.php?xvolno=41_2_9

Kumar, V., & Dahiya, B., (2017). *Smart Economy in Smart Cities, Advances in 21st Century Human Settlements*. Springer Nature Singapore Pte Ltd.. https://doi.org/10.1007/978-981-10-1610-3_1

Lea. R. (2017). *Smart Cities: An Overview of the Technology Trends Driving Smart Cities*. IEEE,

Li, Y., et al., (2021). *Assistive Systems for Mobility in Smart City: Humans and Goods, HCI in Mobility, Transport, and Automotive Systems*. Springer International Publishing, pp. 89–104. https://doi.org/10.1007/978-3-030-78358-7_6

Luusua, A., et al., (2017). No Anthropocentric Design and Smart Cities in the Anthropocene. *Information Technol*, 59, 295–304.

McKinsey & Company (2018). *Smart City Solutions: What Drives Citizen Adoption around the Globe?*. Mckinsey Center for Government, Singapore, pp. 1–60.

Neom (2021). Index @ www.neom.com. https://www.neom.com/index.htm

PR Newswire (2017). HRH the Crown Prince Mohammed bin Salman Announces: NEOM-The Destination for the Future NEOM Incorporates the New Way of Life and the Strategic Crossroads in Trade, Innovation, and Technology with Livability at Its Core. New York, N.Y.: http://www.neom.com/content/pdfs/NEOM_press_release_English_2017.10.24.pdf

Prado, A.L., et al., (2016). Smartness that Matters: Towards a Comprehensive and Human-Centered Characterization of Smart Cities. *Journal of Open Innovation Technology Market and Complexity*, 2(1), December, DOI: 10.1186/s40852-016-0034-z

Rosenbaum, P. L., et al, (2007). Quality of Life and Health-related Quality of Life of Adolescents with Cerebral Palsy. *Developmental Medicine & Child Neurology*, 49(7), 516–521. https://doi.org/10.1111/j.1469-8749.2007.00516.x

Serag El Din, H., (2013). Principles of Urban Quality of Life for A Neighborhood. *Hbrc Journal*, 9(1), 86–92, 2013. https://doi.org/10.1016/j.hbrcj.2013.02.007

Siano, P., et al., (2018). Introducing Smart Cities: A Transdisciplinary Journal on the Science and Technology of Smart Cities. *Smart Cities*, 1(1), 1–3. https://doi.org/10.3390/smartcities1010001

Sorrell, S. (2007). *The Rebound Effect: An Assessment of the Evidence for Economy-Wide Energy Savings from Improved Energy Efficiency*. UK Energy Research Centre (UKERC).

Suma, S., et al., (2018). Automatic Event Detection in Smart Cities Using Big Data Analytics. In Sugimiyanto Suma, Rashid Mehmood, & Aiiad Albeshri (Eds.), *Smart Societies, Infrastructure, Technologies, and Applications. SCITA 2017*. Lecture Notes of the Institute for Computer Sciences, Social Informatics and Telecommunications Engineering (pp. 111–122). Springer, Cham. https://doi.org/10.1007/978-3-319-94180-6_13

Tufek-Memisevic, A., & Stachura E., (2019). Linear Megastructures. An Eccentric Pursuit in Tackling Urban Sustainability Challenges. Srodowisko Mieszkaniowe nr29/2019 PK Kraków, pp. 54–59. https://doi.org/10.4467/25438700SM.19.041.11672

Uzzell, D., & Moser, G., (2006). On the Quality of Life of Environments. *European Review of Applied Psychology*, 56(1), 1–4. https://doi.org/10.1016/j.erap.2005.02.007

Visvizi, A. (2023). Computers and Human Behavior in the Smart City: Issues, Topics, and New Research Directions. *Computers in Human Behavior*, 140, March. https://www.sciencedirect.com/science/article/abs/pii/S0747563222004162

Visvizi, A., & Lytras, M. D. (2018). Rescaling and Refocusing Smart Cities Research: From Mega Cities to Smart Villages, *Journal of Science and Technology Policy Management*, 9(2), 134–145. https://doi.org/10.1108/JSTPM-02-2018-0020

Visvizi, A., & Pérez del Hoyo, R. (2021). Chapter 1 - Sustainable Development Goals (SDGs). in the Smart City: A Tool or an Approach? (An Introduction). *Smart Cities and the UN SDGs*. Elsevier, pp. 1–11. https://doi.org/10.1016/B978-0-323-85151-0.00001-4

Yang, L., et al., (2019). Integrated Design of Transport Infrastructure and Public Spaces Considering Human Behavior: A Review of State-of-the-Art Methods and Tools. *Frontiers of Architectural Research*, 8, 429–453.

Yin, C., et al., (2015). A Literature Survey on Smart Cities. *Science China Information Sciences*, 58, 100102:1–100102:18: https://doi.org/10.1007/s11432-015-5397-4

Yin, C., et al., (2020). *Assistive Systems for Special Needs in Mobility in the Smart Cite.* HCI International, HAL hal-02487688.

5 Unveiling the role of urban discontinuity on equity in public green open spaces

The case of Alexandria, Egypt

Shahira Assem Abdel-Razek and
Sara Mohamed Sabry Zakaria

5.1 Introduction

The continuous evolution of the urban space is a niche topic that is constantly under scrutiny, whether to better digress, analyze, improve, or solely maintain the existing state. This, however, in some cases is not an easy feat, especially if the urban context is in a discontinuous state of development. Coupled with rapid urbanization and urban decay, this has led to inequitable access to services and facilities. The objective of this chapter is to contribute to the existing debate on means to provide a better and more balanced urban environment where access to nature and recreational spaces is not a privilege but a fundamental right for all residents, regardless of their socioeconomic status. More importantly, it offers a new perspective on addressing the equity crisis in public green open spaces, capitalizing on the urban fabric's inherent rifts. By reimagining urban discontinuity as a resource rather than a hindrance, this chapter opens avenues for inclusive urban planning, striving for sustainable city development.

It is in the realm of providing equality, equity, and justice to all that it is without a doubt a right for all to have open public recreational and leisure facilities—areas where individuals can relax, enjoy, exercise, or just socialize. The presence of public green open spaces for all is a right that is often overlooked when developing or discussing new urban planning projects. Governments are at a continuous hassle of providing housing, education, healthcare, and other crucial services that often the right "to play" is not one of the priorities that arise. However, if we are to truly address sustainable development and its effect on the lives of inhabitants, then it is without doubt a very much needed approach to incorporate.

On another note, the growing literature on the evolution of smart cities (Chang and Smith, 2023; Wang and Zhou, 2023) is a continuous and scrutinized view that aims to assist dwellers of urban areas to live fully and up to their real potential through improving their everyday occurrences. This can be realized through many different notions and theories, some of which have been duly researched, while others are still under postulation.

DOI: 10.1201/9781003415930-7

When discussing developmental planning, the set of actions that are carried out is vehemently interlinked and continuous. For a plan of an area to come to light, a master plan is sought out first and then subsections are tackled on a smaller scale. Thus, continuity is recurrently the path taken to achieve sustainability, albeit sometimes, and in contextual settings, discontinuity of urban development may achieve what continuity has not enabled, which may be seen in some cases as an advantage. It is from this last point that this research focuses on offering new creative sights and solutions to the concept of discontinuity as an ardent notion to achieving equity in the leisure and recreational setting.

The continuous evolution of the urban space coupled with rapid urbanization and urban decay has led to inequitable access to services and facilities. Among these facilities is the access to open and public spaces for recreation and leisure, an inherent right that is not fulfilled globally for the urban poor. Through reimagining urban discontinuity as a resource rather than a hindrance, this chapter opens avenues for inclusive urban planning, striving for sustainable city development. The chapter is divided into two main parts: the first part delves into the underlying trends for the scarcity of green open spaces, attributing their absence to urban encroachment and the abandonment of existing areas. The second part is an empirical study on Alexandria city as a case study. Leveraging the spatial gaps and disjunctions within the urban landscape, the authors propose a concept that harnesses these gaps to enhance equitable access to green open spaces.

5.2   Public open spaces as an inherent right to all

Public green open spaces (parks, gardens, and other pieces of land that are covered mostly by the color green; lawns, shrubs, and even moss) (Vidal *et al.*, 2019) are the lungs of any urban area (Jones, 2018). They serve to replenish the amount of oxygen in the atmosphere, improve air quality, decrease air pollution (and, in the long run, help in the reduction of adverse climate change setbacks), and act as places for exercise, recreation, leisure, and social gatherings (Abdel-Razek and Moanis, 2021). Their presence promotes health and well-being (Visvizi *et al.*, 2021), both physically (lowering hypertension, decreasing heart rate, and regulating sleep) and mentally (decreasing anxiety and depression) (Ouf *et al.*, 2021). All in all, green open spaces contribute to the overall enhancement of the quality of life of urban citizens (Francis, 1989).

Access to green public open spaces is set in concrete in the Sustainable Development Goals (SDGs) not only in the sustainable communities' goal (Goal 11), through an intent on making cities "safe, resilient, accessible and sustainable," but also in Goals 5 and 10 dealing with both gender equality and reduced equality between and within countries. The targets for these goals, respectively, are described in Table 5.1 (United Nations, 2015; Visvizi and Pérez-delHoyo, 2021).

Table 5.1 The targets of the SDGs 11, 5, and 10, respectively

Targets	Description
5.c	Adopt and strengthen sound policies and enforceable legislation for the promotion of gender equality and the empowerment of all women and girls at all levels
10.2	By 2030, empower and promote the social, economic and political inclusion of all, irrespective of age, sex, disability, race, ethnicity, origin, religion or economic or other status
10.3	Ensure equal opportunity and reduce inequalities of outcome, including by eliminating discriminatory laws, policies and practices and promoting appropriate legislation, policies and action in this regard
11.7	By 2030, provide universal access to safe, inclusive and accessible, green and public spaces, in particular for women and children, older persons and persons with disabilities

Several initiatives have placed at their heart the focus on making green public spaces a safe, resilient, and accessible place for all, mostly through providing safe zones where mostly women and young girls can go. Initiatives like "make space for girls" (Walker and Clark, 2023), "friends of Rowntree parks girls" (Friends, 2023), and "girlSPARKS" (COPRPS, 2019) all focus on gender-sensitive design to both enable and empower women to make use of green open public spaces for recreational, leisure, social, and sometimes even educational purposes. This came as an afterthought to the research held in Sweden on the demographics of park usage. The study found that before the age of eight, there was no dominant gender in going to play; however, as they aged, parks became predominantly biased to males, due to the types of activities they partook in, skating and football among others (Walker and Clark, 2023). Another study revealed that safety was an issue of concern for women and girls and that the majority felt that it wasn't for them (Baran *et al.*, 2014), as is evident in the 80–20% inhabitants of parks (80 being males and 20 being females) (Safer Parks Consortium, 2023).

The above drove organizations and policymakers to address a new norm, to focus on providing safe access to areas that may not have been on the radar as problematic in the first place, and to provide all youth and women areas where they can carry out their different activities without feeling unwelcome, intimidated, or that they don't belong (WHO, 2017; Housing and Agency, 2023; Safer Parks Consortium, 2023).

Unfortunately, these initiatives have not yet spread to developing and underdeveloped countries; on the contrary, many cities now, due to rapid urbanization and urban decay, may not even have open public spaces in the first place or if they do exist, they are either inaccessible or derelict. The rationale behind their absence is either urban encroachment or abandonment of existing green spaces, resulting in residents going to extra measures to find their own recreation, even if it is on a patch on the side of a highway or main road that has grass. Accessibility to green open spaces, in several cities, is now confined to one of the following types: (a) private areas (sports and social clubs) and (b) gated areas that are open to the public against a fee, ultimately allowing only those with financial means to tread these places.

However, for the sake of equity, financial ability cannot be the basis on which to differentiate between who should or should not access green open spaces; in fact, there shouldn't be any type of differentiation on which to decide who should be allowed to use them. It is with this mind that the researchers focus on coming up with a new concept stemming from the discontinuity in urban space paradigm, focusing on utilizing this rift in the fabric to their advantage.

5.3　Equity in the smart city context

Smart cities as drivers for economic growth and stability have been the go-to concept when discussing any new notion or phenomenon (Kędra *et al.*, 2023; Okafor *et al.*, 2023; Rejeb *et al.*, 2022), coupled with sustainability; they are the sole most researched urban contexts today (Tregua, *et al.*, 2021). However, one cannot survive without the other, regardless of the insufficient literature on the topic, on the grounds of the eminent persistence that the world needs sustainability in any new endeavor it partakes, including and not limited to smart cities and urban development.

The smart city, a much researched and focused upon ideology, is at its very essence a concept that arose to help develop and improve the life of their residents (Visvizi *et al.*, 2018; Khan *et al.*, 2017); whether through introducing new technologies (Montes, 2020), better mobility (Wegener *et al.*, 2017), instant solutions (Lytras *et al.*, 2020), promoting economic growth (Note, 2019), enhancing existing services and facilities (Attaran, *et al.*, 2022), or continuous assessment and improvement (Andrade *et al.*, 2021), at its heart is an inherent notion of the well-being of the individual. The definition of the smart city by Caragliu et al. 2011 sets the well-being of residents as a focus: "when investments in human and social capital and traditional (transport) and modern communication infrastructure fuel sustainable economic growth and a high quality of life, with a wise management of natural resources, through participatory governance" (Caragliu *et al.*, 2011). On the same note, Albino et al. described the smart city as comprising six "urban oriented" parameters: economy, environment, living, people, transportation, and government (Albino, *et al.*, 2015). A third definition emphasized that smart city sustainability and resilience can only be achieved if the well-being of residents is focused upon through information management and technology imperative to the smart city (Lytras and Visvizi, 2020). These works spotlight that the human is at the center of the urban concept. Among this focus is the right to receive equal and fair access to different services.

It is thus not debatable that among this continuous improvement is the need to provide equity to the residents. This equity, while often discussed in the context of digital equity (Calzada and Almirall, 2020; Horrigan, 2019), has now reached further to encompass other forms, especially with the focus on human-centered design and welfare-driven development. Researchers have even gone as far as debating that if the smart city does not provide this type of humanistic design, it may not be that "smart" after all, smart here being people smart rather than data and gadget smart (Mouton *et al.*, 2019; Okafor *et al.*, 2023). It is inevitable then, should the smart context not be found, that prevalence of equity to different circumstances be present, as it will ultimately lead to realizing the true aim of the smart city.

5.4   Continuity and discontinuity in the urban space

Continuity and discontinuity in the urban scope refers to the planning progress in a specific location across a set of dimensions (time, change in authority, planning levels, unseen constraints, and diligence in carrying out the plan). Often, discontinuity is discussed as the abrupt breakage in infrastructure. However, there are many facets to both as is seen in Table 5.2.

The qualms of both the continuous city and the discontinuous city (or what is sometimes referred to as the ruptured) (Solomon, 2018) are as follows:

- The social openness of the city (welcoming and embracing to all citizens and inhabitants)
- Spatially uninterrupted or continuous (buildings and open spaces make way for each other to better blend and divide)
- Continuity is one of evolution: continuous improvement of the existing master plan to accommodate new challenges (resilient)

(Langdon, 2018)

Discontinuity, on the other hand, from the urban planning scope, is often referred to as the discontinuation of urban development and forms resulting in residual areas in the urban context (Guastella, *et al.*, 2019). These areas, although they may seem as areas of neglect, may in turn become areas of hope for a better living, providing wellness and health to the residents of their allocated areas. What may seem as pieces of pure detriment may be an opportunity for urban landscapers to come up with ideas to offer a new scope for leisure, thus enabling derelict or unused land to become something else. Some of the pros of discontinuity in the urban space are

Highlight the complexity and dynamicity of the urban morphogenesis; it may also define the risks and vulnerability of the urban context (Florescu, 2014).

- Create moments of rebirth and metamorphosis of the urban form, in an osmotic relationship between the solid and void of the building and the urban fabric (Carlotti, 2022).
- Challenge the conventional notions of urban and landscape design by offering new creativity patterns and interventions that may aid in coming up with new norms in the urban context.

Table 5.2 Differences between Urban continuity and Urban Discontinuity

Urban continuity	Urban discontinuity
• Refers to a situation where the built environment and various urban elements exhibit a consistent and connected pattern • Implies a smoother transition between different parts of the city, with a sense of cohesion and a more gradual change in physical and functional characteristics • Urban continuity often arises from well-planned urban design, zoning regulations that encourage compatible land uses, and a historical development pattern that emphasizes gradual transitions • Examples of urban continuity could include neighborhoods with consistent architectural styles, mixed-use developments where residential and commercial spaces seamlessly blend, or areas where the transition from one land use to another is gradual and harmonious. In urban planning and design, achieving a balance between discontinuity and continuity is important to create functional and visually pleasing urban environments • The goal is often to promote a sense of community, ease of movement, and efficient land use while preserving the unique characteristics of different neighborhoods or areas within a city	• Refers to a situation with a noticeable break or interruption in the physical, functional, or visual continuity of urban development • Is a situation with gaps or abrupt changes in the built environment, infrastructure, land use, or other urban elements • Often results from factors such as historical development patterns, zoning regulations, topographical features, or economic disparities. Urban discontinuity can lead to areas with starkly different appearances, functions, or social characteristics adjacent to each other, creating a sense of fragmentation within the urban fabric. • Examples of urban discontinuity might include areas with sudden transitions from dense high-rise buildings to open spaces, industrial zones adjacent to residential neighborhoods, or neighborhoods with drastically different architectural styles and infrastructure

Source: Adapted by authors from (Florescu, 2014; Guastella *et al.*, 2019; Carlotti, 2022).

5.5   Continuity vs. discontinuity in the cities of today

The examples of both urban continuity and discontinuity in the urban scope are many; however, some areas may harbor both continuous and discontinuous planning due to many aspects: wealth of the government; policies, laws, rules, and regulations enforced; socioeconomic level of the residents; sustainability model; governance; public perception and involvement; and of course priorities in development. Needless to say, in times of prosperity, human welfare and well-being are put at the focus of any strategy; however, that may diminish when more pressing issues are at hand.

Cities like Paris (France), Portland (Oregon, USA), and Amsterdam (Netherlands) are known for their consistent design. Their urban design is visually appealing and welcoming and focuses on continuity and consistency. On the other hand, cities like Mumbai (India), Rio de Janeiro (Brazil), and Las Vegas (Nevada, USA) are cases of discontinuity, with disparities in development, socioeconomic characteristics, and scenery and landscape. However, there are also cities which possess both continuity and discontinuity and albeit these being mostly in developing countries, the contrast between them is vast. Cities like Cairo and Alexandria in Egypt and Istanbul in Turkey are clear demonstrations of the yoyoing between both continuity and discontinuity, as is evident in the historic places of these cities, in contrast with their present-day development and planning, and the discrepancy in wealth and living conditions.

Against this backdrop, the researchers have chosen the city of Alexandria, Egypt, as a point of focus to highlight the present discontinuities in development, which may act as nodes of a different type of growth (that of the inhabitants living in close quarters).

5.6 Governance in the urban contexts

To better discuss the role of discontinuity in urban space, governance in the urban space should be mentioned. While discontinuity in this proposed case may not be a direct result of governance, or lack of, urban governance may help in guiding urban initiatives to achieve optimum results. Governance, whether carried out by public, private, or non-governmental organizations, helps aid decision makers in how to better deal with a problem (Pierre, 1999). It revolves around providing transparency on any given subject and then being held accountable to the actions and decisions made (Abdel-Razek, 2021).

Ultimately, for the concept at hand, governance would help policymakers in better understanding the reasons behind the discontinuity in the urban space. It will also provide adequate perceptions and insights to the local residents regarding how stakeholders will partake in improving any given problem and the best measures needed to achieve this goal. Equity and inclusiveness are also one of the main components associated with urban governance, consequently transfixing the concept of providing equal and fair opportunities to all levels and citizens of the urban area at hand.

5.7 Empirical study: The case study of Alexandria City

5.7.1 The study area

The city of Alexandria, a historically livid and valuable city renowned for its innovation and creativity, has always been in a continuous state of scrutiny. Whether because of its historic value, its metropolitan state, or its

Mediterranean influences, the city of Alexandria is a cherished and popular area of focus to both scholars and inhabitants. The common demographics of Alexandria are as follows; according to the Central Agency for Public Mobilization and Statistics (CAPMAS), the current population of the Alexandria is now around 5.6 million, with 48% female and the rest male. More importantly, the poverty rate in Egypt in 2020 was 32%, which is roughly 2 million inhabitants of Alexandria alone, and this is the poverty level, not to mention the upper-lower to middle-income strata. The city of Alexandria encompasses eight districts including Montazah, Eastern (Sharq), Middle (Wasat), Western (Gharb), El-Gomrok, El-Agamy, El-Ameriyah, and Borg El-Arab districts. Due to the lack of data for El-Ameriyah and Borg El-Arab districts, the study area (on the city scale) is limited to six districts (Figure 5.1), covering approximately 336.5 km².

The population percentage is distributed as follows: Montazah District has the highest population percentage, followed by Sharq District and then Wasat District, respectively (see Table 5.3). This population is based on the latest census carried out by CAPMAS in 2016; since then, the population of the above districts has increased incrementally by approximately 1.9% per year.

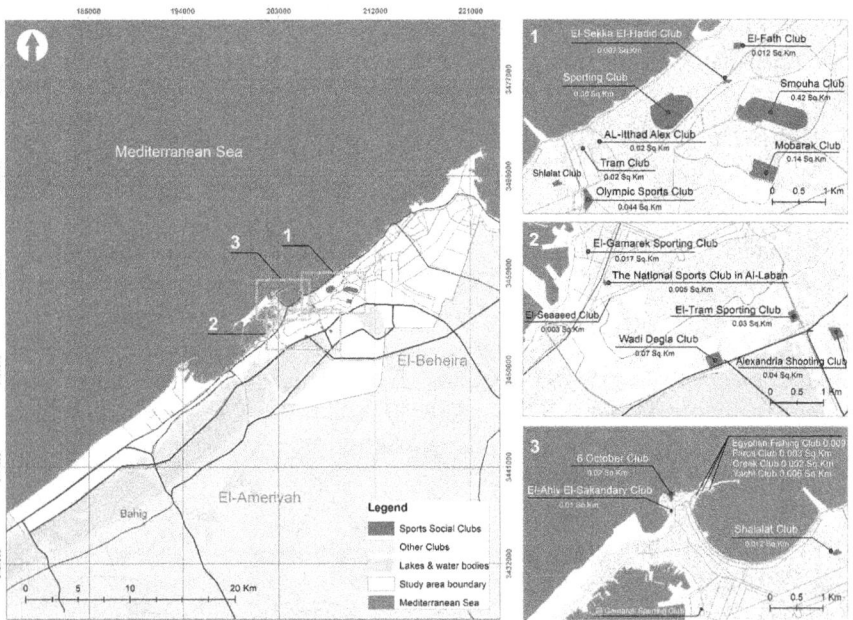

Figure 5.1 Mapping between sports and social clubs in Alexandria and other types of clubs.

Source: Authors.

Table 5.3 The area and population of the study area districts

	District	Area (sq. km)	Population
1	Montazah	135.81	1,585,572
2	Sharq	50.87	1,158,822
3	Wasat	29.60	543,405
4	Gharb	73.88	356,613
5	Gomrok	4.68	156,780
6	Agamy	95.64	472,721

Source: Central Agency for Public Mobilization and Statistics (*CAPMAS*, 2023).

5.7.2 The economic profile of Alexandria

Alexandria is Egypt's second largest city, with a population of approximately 5.6 million, 47% of which is under the age of 25 and is expected to increase to 6.8 million by the year 2032. The percentage of unemployment was around 12% in 2021 (Galal, 2023), with 54% of these between the ages 15 and 40 (Galal, 2022). The estimated percentage of citizens living in low socioeconomic levels (whether formal or informal) is around 25%, which is to say that these families cannot afford to pay to go to places of recreation or leisure and need open green public spaces that are free and at the same time inviting, welcoming, and accommodating to their needs.

The income poverty in Egypt in 2018 (according to the last national census) was around 32% living in middle- to low-income households (Armanious, 2018); whether formal or informal housing, as such, many of these households do not possess access to basic services and primary amenities, much less to areas of recreation, or leisure. This can lead to not only physical and mental problems, but also problems associated with the increase in social divide and the frustration that may arise from it.

5.7.3 Methodology, tools, and data sources

The methodology used in this study is based on a mixed approach combining thorough theoretical analytical and empirical studies. The aim is to gain a comprehensive understanding of the present urban discontinuities in development, focusing on public green open spaces and utilizing this rift in the fabric to their advantage. In the theoretical analytical part, a review of the existing literature on discontinuity in urban space and equity in the smart city context is carried out. The most relevant studies are reviewed to identify the main theories and findings related to the current study. Second, an empirical study is conducted using qualitative research which includes mapping technique and visual analysis. The study is carried out in Alexandria city context as a case study.

The supporting data used for thematic mapping are added to ArcMap 10.7.1. The shapefiles of city administrative boundaries, the buildings, the land uses, and road network are employed in a geodatabase for Alexandria city. The data required for the analysis are acquired from different sources such as the CAPMAS, Alexandria Passenger Transportation Authority (APTA), and General Organization for Physical Planning (GOPP). The base map (i.e., satellite image) is used as a reference map to make additional observations to support the thematic maps' preparation. Once the data have been collected, they are analyzed and interpreted. Finally, the results and analyses obtained from the literature review and the empirical study are combined to offer a comprehensive view to propose a strategy or possible intervention that harnesses these gaps to enhance equitable access to public green open spaces.

5.8   Analysis of the proposed study area

5.8.1   Analysis and interpretation

The first step of the adopted methodology begins with mapping the different features that are essential to build the analysis for the current study and these features include (1) main sports and social clubs (Figure 5.1), (2) public gardens and open spaces (Figure 5.2), (3) main gardens that are gated and open to the public against a fee (Figure 5.3). For the second step, during the mapping process, the authors had to compare the public gardens and open spaces with the built-up areas among the study area. Using building footprints, the thematic maps can show the lack of public green open spaces relative to the total area of the study area and the condensed surrounding urban fabric. For the third step, the authors took a closer look at the vacant lots and abandoned lands that could have potential for possible interventions.

As can be seen in Figure 5.1, there are nine main sports and social clubs in Alexandria: these clubs are privately owned clubs with the number of members combined not reaching a million of the population, meaning that an approximate of around 5 million inhabitants do not have direct access to any of these social areas. Most of those who are not members are either not interested in attaining membership or simply cannot afford it, with the majority being the latter. On the other hand, the areas of open public spaces that are open, free, and readily available to the public are less than 1% of the total area of the city, as is evident in Figure 5.2; the total area of green places open to the public free of charge is 1.30 km², contributing to 0.39% of the total area of Alexandria city. This is less than the 15% norm identified by the UN-Habitat as being the minimum area for public green open spaces.

Regarding the main gardens that are open to the public for a fee, the city of Alexandria harbors three main gardens: El-Montazah Gardens, Antoniadis Gardens, and El-Shalalat Gardens. El-Montazah and Antoniadis are both gated gardens and entrance is against a ticket per person and vehicle (an exorbitant fee that is not within the means of a family of four), while El-Shalalat is free of charge, but has gone to waste and has a reputation of being a place that is not safe; mostly, only males go there. These three gardens are approximately 2 km² in area (Figure 5.3).

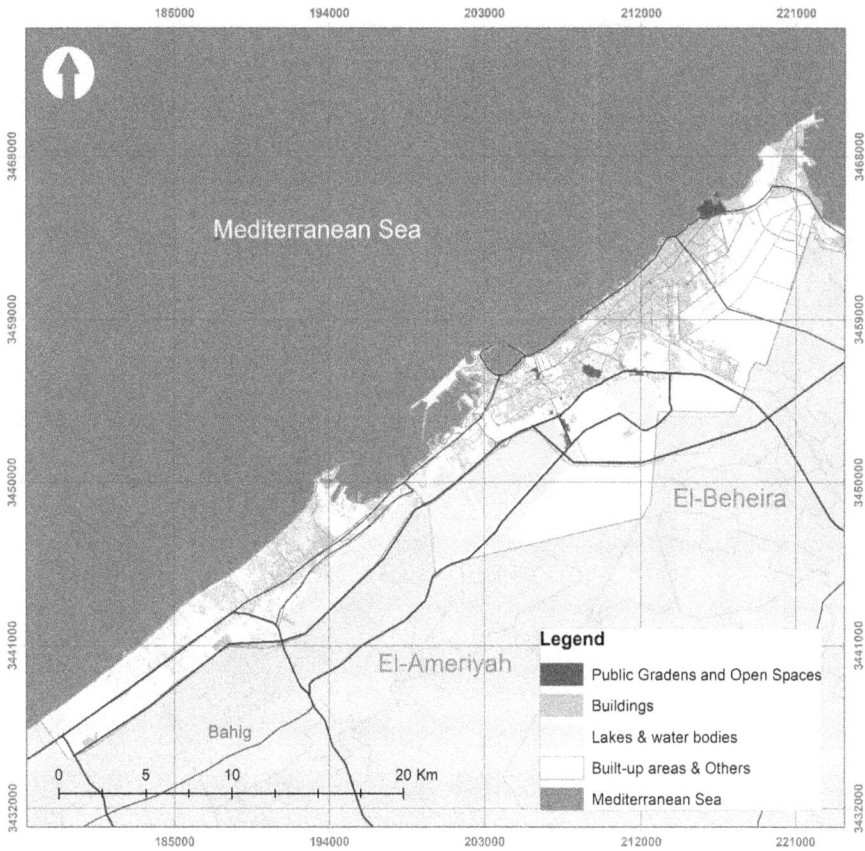

Figure 5.2 Map shows the ratio between the public green open spaces and the other built-up areas in Alexandria City.

Source: Researchers.

Although the percentage of public green open space areas available for recreational, leisure, and social purposes is insufficient, there are a number of vacant lots and abandoned land that have continuous planning on all their sides but apparently been neglected. These lands are all public governmental lands, some of which used to be old country-owned industrial sites and factories. As is evident from Figures 5.4 and 5.5, these abandoned lands are located within areas of high urban density, meaning that their location is in areas of either high poverty levels or low socioeconomic status.

It should be noted that the nine proposals of lands with total area 3 km^2 (Table 5.4) encompass a diverse selection within Alexandria that suits their population distribution. Among these options, there are three pieces of land situated in Montazah district, another three in the Sharq district, two more in Agamy district, and a single piece of land proposed in the heart of the Wasat

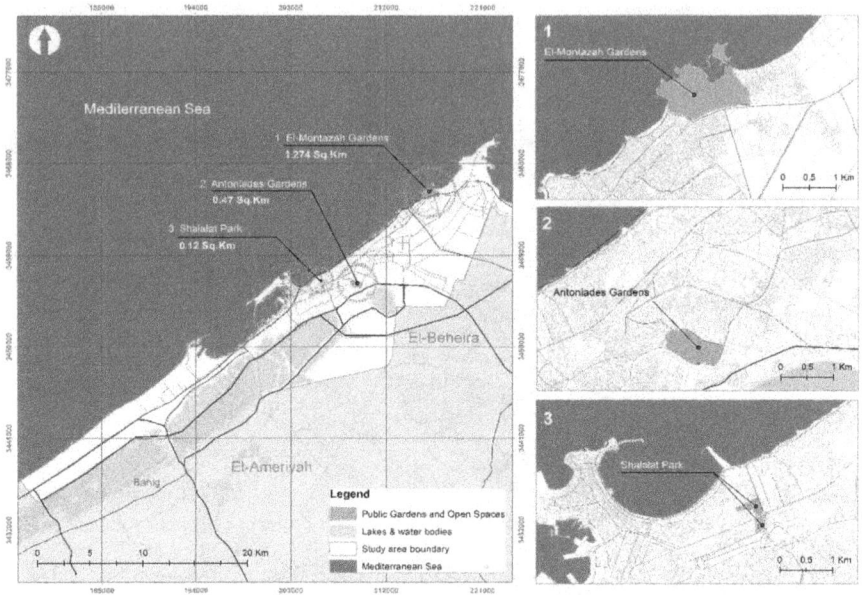

Figure 5.3 Existing main gardens in Alexandria city; El-Montazah, Antoniadis, and El-Shalalat.

Source: Researchers.

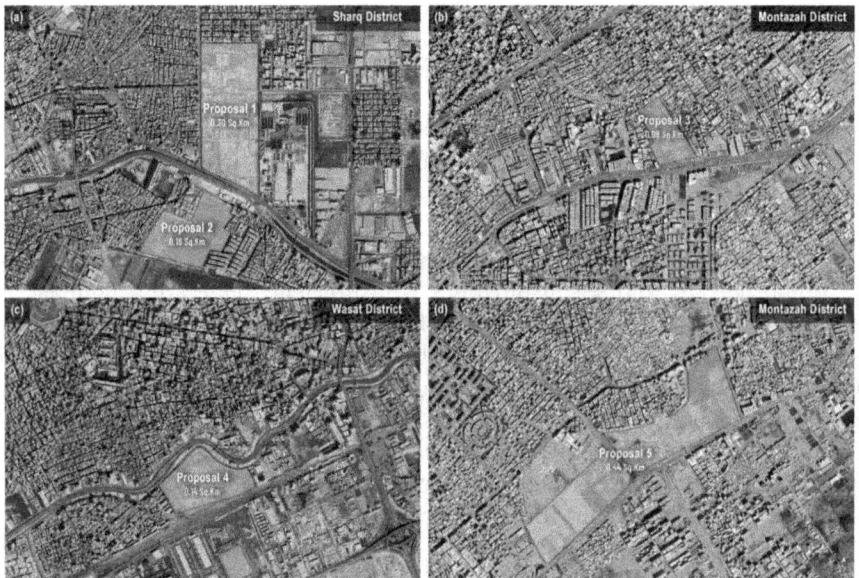

Figure 5.4 The satellite images for the proposed lands 1–5 show the surrounding high urban densities.

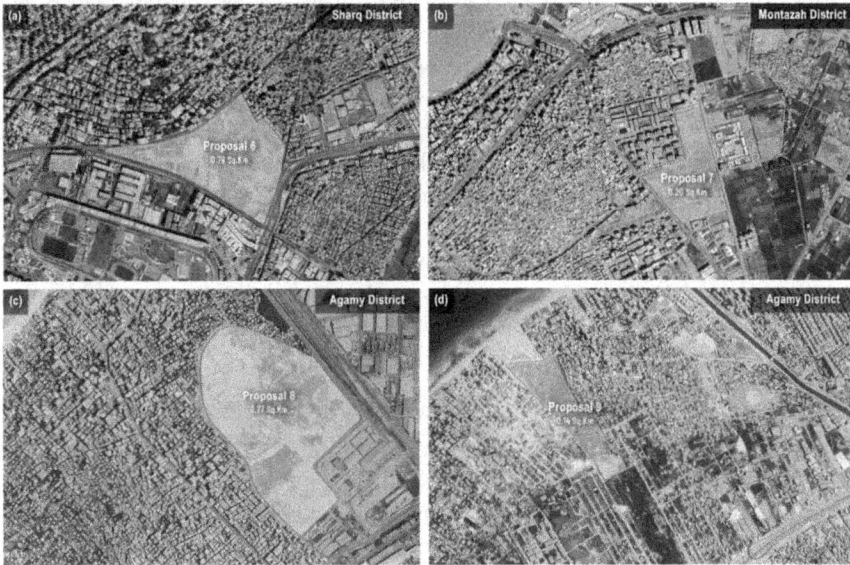

Figure 5.5 The satellite images for the proposed lands 6–9 show the surrounding high urban densities.

Table 5.4 Size and previous vocation of proposed land

Land Number	Area (sq. km)	District	Current ownership/use
Proposal 1	0.30	Sharq	El-Seyouf company for clothes and textiles or Sabahi company (abandoned)
Proposal 2	0.16	Sharq	Vacant land
Proposal 3	0.08	Montazah	Eastern Linen and Cotton Company (abandoned)
Proposal 4	0.14	Wasat	Vacant land
Proposal 5	0.44	Montazah	Vacant land
Proposal 6	0.79	Sharq	Vacant land (coppwhoer company ownership)
Proposal 7	0.20	Montazah	Vacant land
Proposal 8	0.77	Agamy	Vacant land
Proposal 9	0.14	Agamy	Vacant land

district. Notably, there is no land proposed in Gomrok district, owing to its status as the historical nucleus of the city. Moreover, Gomrok district, with the smallest area and population in Alexandria, maintains its charm through its enduring adherence to the old city's timeless planning principles.

Understandably, six out of the nine lands proposed for development lie in both Sharq and Montazah districts, located at the east end of the city; these are the two most populous districts, as is seen in the overall populations, in Table 5.3, of the city but their proximity to the sea and the city center sustains their popularity. Since there is a little more room for development in these districts than in the old core districts, they continue to be very popular among growing families (GOPP, 2014). These proposed lands vary in areas where the proposed interventions could be a large public park (for large areas). Generally, there are a number of other vacant lands (smaller in size) that could be useful for creating 'pocket parks,' green spaces which can create new public spaces without the need for large-scale redevelopment; however, most of these areas are privately owned lands and will need country-wide initiatives and enforcement to be able to transfer ownership of these lands into public trust, which will take time and money, not to mention that in the meanwhile they may turn into built-up areas.

5.8.2   *Advantages of using derelict, vacant, and abandoned pieces of land*

To better achieve governance in the governorate of Alexandria, consequent local governments need to work in sync and allocate funds to improve the quality of life of the residents of the city. This includes utilizing present vacant and abandoned areas as equity value designed to better embrace the needs of the general public, meaning that instead of turning these lands into built projects, they can be better incorporated as open public spaces that will not only act as lungs to the community, thus mitigating the effects of climate change and improving air quality, but also increase the surrounding land value. For instance, as seen in Figure 5.6, there are proposed interventions for lands located in Sharq, Agamy, and Montazah districts to be upgraded and developed to establish a public green open space. For the sake of resilient planning and design, the intervention could also be a multifunctional green open space that maintains mitigation measures. The present state of derelict ensues that during rainstorms, rainwater causes flooding, ponding, and damage to paved areas and streets; it represents an urgent public health hazard. These lands could be used as public urban space as well as alternative energy creation, rainwater management, or flood retention corridors which can be adaptable to different needs. Ultimately, these types of projects can help to achieve the following advantages, financially, environmentally, and socially:

1 Create high-revenue investment opportunities.
2 Provide job opportunities and income-generating activities.
3 Empower women and children.
4 Establish green lungs to support microclimate 'urban cooling.'
5 Reduce air pollution.
6 Maintain mitigation measures through adaptive urban design of these spaces.
7 Enhance the quality of life of residents.
8 Provide recreational, leisure, and social open space for all.
9 Increase the surrounding land value.
10 Decrease crime rates and enhance safety.

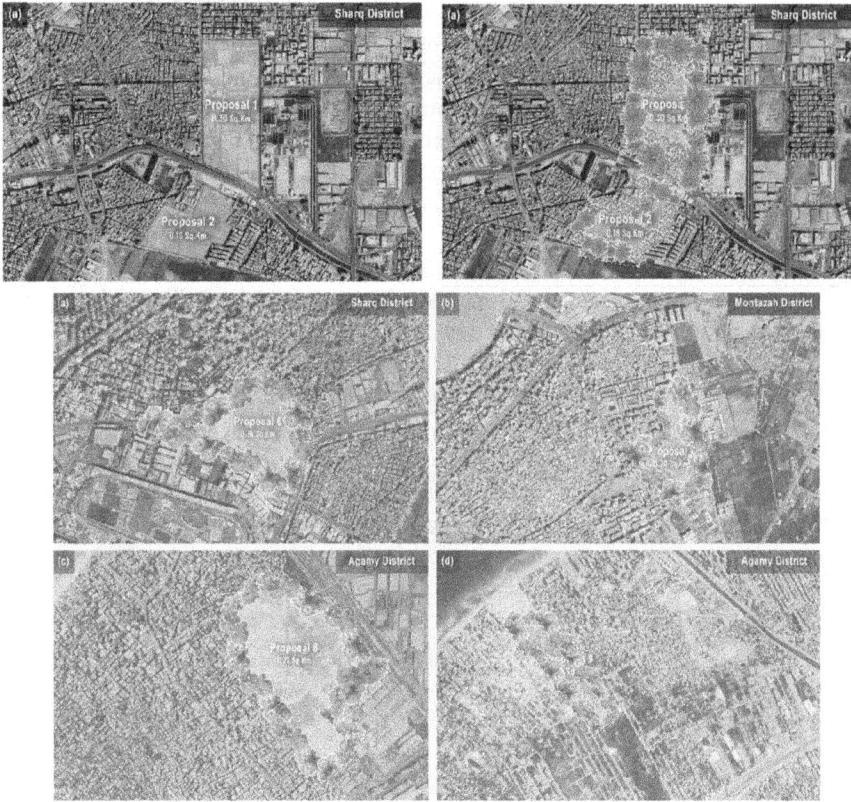

Figure 5.6 Schematic proposal of the green area and how it will change the visual perception of the area.

5.8.3 *Provision for green open spaces: Concept and principles*

From the previous, managing, upgrading, and developing existing urban areas need equal attention from urban designers, urban planners, and decision makers as building new projects. There is also a need to formulate context-specific solutions and the level of interventions needed to maximize the spatial potential of these proposed lands. Urban designers should work on detailed planning principles and design components to include discontinuous planning areas into current and future plans, especially for segregated and marginalized groups. Utilizing vacant and abandoned lands may be the start of integrating derelict areas into the overall quality of life balance of citizens; this can only be achieved through a detail-oriented plan that assesses, evaluates, plans, implements, monitors, and revisits shortcomings and is adopted by all stakeholders.

This concept at hand needs a set of guiding principles for utility of lands to promote creation of green open spaces on the premises of discontinuity:

- An overall assessment of the different lands available, their ownership, status, whether they lie in any planning proposals, cost of development and maintenance, and number of beneficiaries in the long run.
- Lands should be vacant, with minimum built-up area; however, if a building does exist, retrofitting it into a social or cultural zone is advised.
- Lands should be put in a trust owned by the government for the public to attest their survival and continuation.
- Areas should be made accessible through different urban modes to link them with the existing urban fabric.
- Community involvement and participation should be utilized during the decision-making process to empower and enable families to take responsibility and action.
- Associations and agencies should be chartered for administrating, managing, and running the developed area.
- Funds should be allocated to ensure continuity through setting up an urban land management entity.
- Zoning codes, laws, regulations, and enforcement practices should include and embed new amendments for green open spaces mandates.
- Incentives should be given to developers who integrate open public spaces into their projects.

With the empowerment of the community and once they realize the advantages and impacts of similar projects on their quality of life, well-being, and health, social resentment and hostility may well decrease. The public will start to feel that they are responsible for this area and, especially if community participation and involvement is utilized, for what is happening in their neighborhoods and regain confidence.

5.9   Conclusion

The main focus of any developmental project are the inhabitants of that area; whether this project aims to develop, upgrade, or maintain living conditions, it is inevitable that well-being be promoted above all else. Well-being not only encompasses the physical aspects of the person's life, but also his psychological and social state and providing equity to areas that may not seem within reach to the urban poor while, at the same time, retaining their dignity is not an easy feat. In this regard, continuity and discontinuity in urban space may play a vital role, especially, if this discontinuity is turned from something that is unfavorable to something that is highlighted and delved upon: public open spaces. This chapter offers a novel concept focusing on addressing the equity crisis in public green open spaces, capitalizing on the urban fabric's inherent rifts. The authors propose that urban discontinuity may be looked upon as a resource rather than a hindrance and propose avenues for inclusive urban planning striving for sustainable city development.

While discontinuity in the urban planning process may seem like a drawback, utilizing spaces that have not been developed or places that are abandoned for projects like inclusive parks and other recreational and leisure areas may offer the required initiative to provide equal opportunities to all strata of the community: a place where people can go, without necessarily paying for entrance to spend time, exercise, and socialize and without being continuously subjected to financial anxiety. The mapping of areas in places like the city of Alexandria, Egypt, has shown that there are several vacant lands that have not yet been included in the developmental projects. These vacant lots have been abandoned for some time or are a remiss of planning initiatives (thus discontinuity). The current research proposes that instead of including these areas into developmental zones as buildings, they can become open public spaces for all. In this chapter, the researchers focused on turning derelict or vacant land that is owned by the government to favorable public open spaces to the general public. A criterion for the choice of the land was proposed and several lots were chosen with images of suggested before- and after-shots. These areas of land were chosen in high-density urban areas that clearly lack any form of public open spaces, with low to lower-middle socioeconomic levels.

References

Abdel-Razek, S. A. (2021) 'Governance and SDGs in Smart Cities Context', in *Smart Cities and the un SDGs*. doi: 10.1016/b978-0-323-85151-0.00005-1

Abdel-Razek, S. A. and Moanis, Y. (2021) 'Resilient Urban Open Public Spaces in During the Covid-19 Pandemic', in Visvizi, A., Troisi, O., and Saeedi, K. (eds) *Research and Innovation Forum 2021*. Cham: Springer International Publishing, pp. 253–268.

Albino, V., Berardi, U. and Dangelico, R. M. (2015) 'Smart Cities: Definitions, Dimensions, Performance, and Initiatives', *Journal of Urban Technology*, 22(1), pp. 21–23. doi: 10.1080/10630732.2014.942092

Andrade, R.O.; Yoo, S.G.; Tello Oquendo, L.; Ortiz-Garcés, I. (2021) 'Chapter 12 - Cybersecurity, Sustainability, and Resilience Capabilities of a Smart City', in Visvizi, A. and Pérez del Hoyo, R. B. T. (eds). *Smart Cities and the un SDGs*, Elsevier, pp. 181–193. https://doi.org/10.1016/B978-0-323-85151-0.00012-9

Armanious, D. M. (2018) 'Accelerating Global Actions for a World Without Poverty', *Inter-agency Expert Group Meeting on Implementation of the Third United Nations Decade for the Eradication of Poverty (2018-2027)*, (April), pp. 1–22.

Attaran, H., Kheibari, N. and Bahrepour, D. (2022) 'Toward Integrated Smart City: A New Model for Implementation and Design Challenges', *GeoJournal*. doi: 10.1007/s10708-021-10560-w

Baran, P. K., Smith, W. R., Moore, R. C., Floyd, M. F., Bocarro, J. N., Cosco, N. G. and Danninger, T. M. (2014). Park Use Among Youth and Adults: Examination of Individual, Social, and Urban Form Factors. *Environment and Behavior*, 46(6), 768–800. https://doi.org/10.1177/0013916512470134

Calzada, I. and Almirall, E. (2020) 'Data Ecosystems for Protecting European Citizens' Digital Rights', *Transforming Government People Process and Policy*, 14, pp. 133–147. doi: 10.1108/TG-03-2020-0047

CAPMAS (2023). Available at: https://www.capmas.gov.eg/ (Accessed: 20 September 2023).

Caragliu, A., Del Bo, C. and Nijkamp, P. (2011) 'Smart Cities in Europe', *Journal of Urban Technology*, 18(2), pp. 65–82. doi: 10.1080/10630732.2011.601117

Carlotti, P. (2022) 'Thinking Places: Towards Application of Place-Based GIS in Urban Morphology', in *Urban Form and the Sustainable and Prosperous Cities; XXVIII International Seminar on Urban Form ISUF2021*, pp. 72–78.

Chang, S. and Smith, M. K. (2023) 'Residents' Quality of Life in Smart Cities: A Systematic Literature Review', *Land*, 12(4). doi: 10.3390/land12040876

Coprps, G. (2019) *Igniting Potential through Girl-Centered Design*. Available at: https://girlsparks.org/ (Accessed: 15 August 2023).

Florescu, T. (2014) 'Continuity and Discontinuity of Urban Form—The Issue of Risk', in Bostenaru Dan, M., Armas, I., and Goretti, A. (eds). *Earthquake Hazard Impact and Urban Planning*. Springer Netherlands, pp. 195–211. doi: 10.1007/978-94-007-7981-5_10

Francis, M. (1989) 'The Urban Garden as Public Space', *Places*, 6(1), pp. 52–59.

Friends, T. (2023) 'Friends of Rowntree Park', *Economist*, 395(8680). Available at: https://rowntreepark.org.uk/make-space-for-girls-older-girls-and-rowntree-park/

Galal, S. (2022) *Unemployment Rate in Egypt from 1st Quarter 2021 to 3rd Quarter 2021, by Age Group*. Available at: https://www.statista.com/statistics/1297850/quarterly-unemployment-rate-in-egypt-by-age-group/ (Accessed: 1 September 2023)

Galal, S. (2023) *Unemployment Rate in Egypt in 2021, by Governorate*. Available at: https://www.statista.com/statistics/1297880/unemployment-rate-in-egypt-by-governorate/ (Accessed: 1 September 2023).

General Organization for Physical Planning (2014) Strategic Urban Plan (SUP) Alexandria 2032 – Phase one: Detailed City Profile, Volume 2a, (June 2014)

Guastella, G., Oueslati, W. and Pareglio, S. (2019) 'Patterns of Urban Spatial Expansion in European Cities', *Sustainability (Switzerland)*, 11(8), pp. 1–15. doi: 10.3390/su11082247

Horrigan, J. B. (2019) 'Smart Cities and Digital Equity: Some Cities Have Found Ways to Pursue Both, But Federal Policy May Hinder Progress', (June), p. 11.

Housing, T. and Agency, R. (2023) 'Inclusive Spaces and Places for Girls and Young People an introduction for Local Government', (June).

Jones, K. (2018) '"The Lungs of the City": Green Space, Public Health and Bodily Metaphor in the Landscape of Urban Park History', *Environment and History*, 24, pp. 39–58. doi: 10.3197/096734018X15137949591837

Kędra, A., Maleszyk, P. and Visvizi, A. (2023) 'Engaging Citizens in Land Use Policy in the Smart City Context', *Land Use Policy*, 129, p. 106649. doi: 10.1016/j.landusepol.2023.106649

Khan, M. S., Woo, M., Nam, K. and Chathoth, P. K. (2017) 'Smart City and Smart Tourism: A Case of Dubai', *Sustainability (Switzerland)*, 9(12). doi: 10.3390/su9122279

Langdon, P. (2018) *The Continuous City versus the Ruptured City*. Available at: https://www.cnu.org/publicsquare/2018/12/04/%E2%80%98continuous-city%E2%80%99-versus-%E2%80%98ruptured-city%E2%80%99 (Accessed: 30 August 2023).

Lytras, M. and Visvizi, A. (2020) 'Information Management as a Dual-Purpose Process in the Smart City: Collecting, Managing and Utilizing Information', *International Journal of Information Management*, 56, p. 102224. doi: 10.1016/j.ijinfomgt.2020.102224

Lytras, M. D., Visvizi, A., Chopdar, P. K., Sarirete, A. and Alhalabi, W. (2020) 'Information Management in Smart Cities: Turning End Users' Views into Multi-Item Scale Development, Validation, and Policy-Making Recommendations', *International Journal of Information Management*. doi: 10.1016/j.ijinfomgt.2020.102146

Montes, J. (2020) 'A Historical View of Smart Cities: Definitions, Features and Tipping Points', *SSRN Electronic Journal*. doi: 10.2139/ssrn.3637617

Mouton, M., Ducey, A., Green, J., Hardcastle, L., Hoffman, S., Leslie, M. and Rock, M. (2019) 'Towards "Smart Cities" as "Healthy Cities"', *Canadian Journal of Public Health / Revue Canadienne de Santé Publique*, 110(3), pp. 331–334. Available at: https://www.jstor.org/stable/27174068

Note, I. (2019) '1st OECD Roundtable on Smart Cities and Inclusive Growth', (July).

Okafor, C. C., Aigbavboa, C. and Thwala, W. D. (2023) 'A Bibliometric Evaluation and Critical Review of the Smart City Concept – Making a Case for Social Equity', *Journal of Science and Technology Policy Management*, 14(3), pp. 487–510. doi: 10.1108/JSTPM-06-2020-0098

Ouf, T. A., Makram, A. and Abdel Razek, S. A. (2021) 'Design Indicators Based on Nature and Social Interactions to Enhance Wellness for Patients in Healthcare Facilities', in Trapani, F. et al. (eds) *Advanced Studies in Efficient Environmental Design and City Planning*. Cham: Springer International Publishing, pp. 449–461.

Pierre, J. (1999) 'Models of Urban Governance: The Institutional Dimension of Urban Politics', *Urban Affairs Review*, 34(3), pp. 372–396. doi: 10.1177/10780879922183988

Rejeb, A., Rejeb, K., Abdollahi, A., Keogh, J. G., Zailani, S. and Iranmanesh, M. (2022) 'Smart City Research: A Bibliometric and Main Path Analysis', *Journal of Data, Information and Management*, 4(3), pp. 343–370. doi: 10.1007/s42488-022-00084-4

Safer Parks Consortium (2023) 'Safer Parks: Improving Access for Women and Girls'. doi: 10.48785/100/151

Solomon, D. (2018) *Housing and the City: Love vs. Hope*. Schiffer.

Tregua, M., D'Auria, A. and Bifulco, F. (2021) '3 - Sustainability in Smart Cities: Merging Theory and Practice', in Visvizi, A. and Pérez del Hoyo, R. B. T. (eds). *Smart Cities and the un SDGs*. Elsevier, pp. 29–44. doi: 10.1016/B978-0-323-85151-0.00003-8

United Nations (2015) 'The 2030 Agenda for Sustainable Development's 17 Sustainable Development Goals (SDGs)', pp. 10–15. Available at: https://sustainabledevelopment. un.org/content/documents/21252030 Agenda for Sustainable Development web.pdf.

Vidal, D. G., Barros, N. and Maia, R. L. (2019) 'Public and Green Spaces in the Context of Sustainable Development BT', in Leal Filho, W. et al. (eds). *Sustainable Cities and Communities*. Springer International Publishing, pp. 1–9. doi: 10.1007/978-3-319-71061-7_79-1

Visvizi, A., Abdel-Razek, S.A., Wosiek, R. and Malik, R. (2021) 'Conceptualizing Walking and Walkability in the Smart City through a Model Composite w2 Smart City Utility Index', *Energies*, 14(23). doi: 10.3390/en14238193

Visvizi, A., Lytras, M., Damiani, E. and Mathkour, H. (2018). 'Policy Making for Smart Cities: Innovation and Social Inclusive Economic Growth for Sustainability', *Journal of Science and Technology Policy Management*, 9(2), pp. 126–133. doi: 10.1108/jstpm-07-2018-079

Visvizi, A. and Pérez-delHoyo, R. (2021) *Smart Cities and the UN SDGs*.

Walker, S. and Clark, I. (2023) 'Make Space for Girls the Research Background 2023', (January). Available at: https://assets.website-files.com/6398afa2ae5518732f04f791/6 3f60a5a2a28c570b35ce1b5_Make Space for Girls - Research Draft.pdf.

Wang, M. and Zhou, T. (2023) 'Does Smart City Implementation Improve the Subjective Quality of Life? Evidence from China', *Technology in Society*, 72, p. 102161. doi: 10.1016/j.techsoc.2022.102161

Wegener, S., Raser, E., Gaupp-Berghausen, M., Anaya Boig, E., Nazelle, A., Eriksson, U., Gerike, R., Horvath, I., Iacorossi, F., Rothballer, C., Sanchez, J., Int Panis, L., Kahlmeier, S., Nieuwenhuijsen, M., Mueller, N. and Rojas, D. (2017) 'Active Mobility – The New Health Trend in Smart Cities, or Even More?', *Conference: Proceedings of 22nd International Conference on Urban Planning, Regional Development and Information Society*, (January 2019), pp. 21–30.

WHO (2017) Urban Green Spaces: A Brief for Action (2017) 'Urban Green Spaces: A Brief for Action', *Regional Office for Europe*, p. 24. Available at: http://www.euro. who.int/__data/assets/pdf_file/0010/342289/Urban-Green-Spaces_EN_WHO_web. pdf?ua=1

Part II

Territory, scale, inclusion, and participation in the smart city debate

6 Toward the metaverse. Smartification of public space management

What do we learn from smart cities in the EU?

Tomasz Pilewicz

6.1 Introduction

Smartification of public space management is an important topic for several types of local development stakeholders (Lv et al., 2022). For the purpose of the discussion in this chapter, public space is understood as external public spaces and refers to public recreational areas, transportation routes, parking zones, and excludes, e.g., public office spaces, libraries, or museums. Versatile aspects of smartification of public space management may advance local authorities in urban planning and sustaining local development processes (Visvizi & Lutras, 2019). Second, in the context of growing importance of empowering local citizens and enterprises and involving them in decision-making, it may be one of the key enablers in this context, as the new solutions in smartification domain enable improved allocation of resources and increase ownership resulting from a direct involvement into the process (Lv et al., 2022). The issue is also interesting in local development theories context, including local and regional innovation systems, learning locations and regions, smart specialization of locations and regions, and their resilience to development shocks (Ajobiewe, 2020). Smart organizations and smart cities have been explored for at least three decades and aspects related to public space management advance understanding of smart growth, which smart cities are oriented toward.

The objective of this chapter is to explore the direction of smartification processes, and, in this context, the role of smartification process-related solutions and practices in the public sector domain. Various theoretical perspectives and empirical findings have already addressed the smart city domain (Bell et al., 2017, Allam et al., 2022, Bibri et al., 2022). However, there were no perspectives and findings dedicated specifically to advancement and evolution of a public space management through technology, including rapidly proliferating in recent years the concept of the so-called digital twins and metaverse.

One of the recommended research avenues relating to smart organizations refers to how smart cities through conscious public space management can reinforce sustainability and resilience to shocks of local development processes (Godlewska-Majkowska et al., 2023). This puzzle needs to be addressed as in

DOI: 10.1201/9781003415930-9

contemporary post-pandemic global slowdown; local authorities are actively looking for solutions contributing to the acceleration of development process through engaged, conscious, and well-informed stakeholders, such as citizens and enterprises.

The research presented in the chapter advances the understanding of smartification of public space management through explanation of what smart cities and smart organizations in public sector are about, what smartification in public sector is, and how it evolves. Theoretical part of the chapter finishes with research questions (RQs) inspired by the subject-matter literature referring to the types of solutions and practices used by smart cities to advance public space management domain, benefits and risks related to the usage of solutions and practices used by smart cities in public space management domain, and the direction of smartification process of public space management in relation to solutions and practices used by smart cities. After empirical research methodology design on smartification of public space management, the findings of electronic audit of official Internet portals, and surveys with representatives of smart cities in the EU, conclusions follow.

6.2   Smart cities and smart organizations in public sector

The promotion of smart cities and smart municipalities in the last decade in the EU enriched the discussion on smart organizations in the public sector and the role of local administrative units (LAUs) in perpetuating local development processes and resilience to development shocks. Smart cities are described through the specific ability of the usage of information and communication technology to increase the quality of living of their inhabitants. The dominance of technology solutions in smart cities is discussed to advance urban planning, management, and governance. The set of technology solutions relates to, e.g., collection, processing, and leveraging of data from municipal networks and media installations, parking spaces, traffic lights, water supply systems, sewers, or public monitoring systems (Jiang et al., 2018).

The initial focus of smart city initiatives investigated among others by the Organization for Economic Cooperation and Development (OECD) relates to digital information and communication technology that are enhancing the efficiency of urban services planning and delivery. The evolution of scholar discourse and the role of smart cities allowed the inclusion of the effects they have on the people, the environment, and the local development models (OECD, 2020). Smart city definition in the EU introduced by the European Commission (EC) accentuates their higher efficiency of leveraging traditional networks and services through digital and telecommunication technology for the benefit of inhabitants and businesses (European Commission, 2014).

As the smart city concept is unceasingly evolving and a subject of debate, a smart organization as a broader term contributes to its understanding. Smart organizations in private sector and enterprises have been a more popular subject of research (Chavarría-Barrientos et al., 2015; Godlewska-Majkowska &

Pilewicz, 2020; Adamik & Sikora-Fernandez, 2021) than smart organizations in the public sector, which makes the latter interesting to explore. The public sector consists of entities owned and reporting to the state, LAUs, or self-government entities, where most of the ownership and equity can be attributed to public entities. Public sector and public sector organizations at the country level are often described by legal regulations stating their purpose, obligatory tasks, and budgeting.

By its nature, the public sector is mainly oriented toward the satisfaction of needs and legal entitlements of stakeholders it serves, less the economic profits; hence, in the chapter, attention is paid on public sector organizations responsible for governance over LAUs and their responsibility for the provision of defined public goods and services. A smart organization in the public sector is directed toward better satisfaction of the needs of local development stakeholders it serves, such as inhabitants, investors, and entrepreneurs. Scholars also characterize the ability of smart organizations in the public sector to manage knowledge and carry out public tasks, by being creative and innovative in problem solving, capable of forecasting social needs, and flexible in responding to them (Sikora-Fernandez, 2013).

One of the definitions of smart organization in the public sector defines it as an organizational way of working within the public sector units that effectively manage information, knowledge, communication, relations with partners, leverage technology to deliver upon public tasks, and dynamize local development processes to achieve and maintain competitive advantages (Godlewska-Majkowska & Komor, 2019). This definition due to the combination of aspects ranging from organization, data collection, communication with local development stakeholders, and purposeful usage of technology solutions has been set as a key reference point in the empirical research design detailed in the further parts of the chapter.

Smart organizations in the public sector are very often referred to as smart cities and their respective measurement methods are described in subject-matter literature. One of the most popular measurement approaches of smart cities has been developed by the IMD-SUTD (IMD, 2022) in the form of the Smart City Index (SCI). It assesses the perceptions of city residents gathered within a survey on issues related to two pillars – structures and technology applications available to them in their city (1) and existing city infrastructure (2). Each pillar is evaluated over five key areas: health and safety, mobility, activities, opportunities, and governance. In 2022 edition of the SCI, 13 out of 31 cities from Europe were ranked and analyzed (Amsterdam, Barcelona, Berlin, Brussels, Chicago, Copenhagen, Dublin, Helsinki, Lisbon, London, Madrid, Stockholm, and Vienna). These approaches to measuring smart city performance, including SCI, are often finalized with rankings which appear on an annual basis and enable to identify cities classified as smart ones. In smart cities research, it may foster purposive research sampling and enable better understanding of ways of working of smart cities recognized in rankings and being perceived as reference organizations to other organizations.

6.3 The smartification process in public space management domain

Scholars indicate that the advancement in technology enables smartification of processes understood as increase of their efficiency through management of information and knowledge and communication of stakeholders involved in their delivery (Bibri et al., 2022). The advancement in information and communication technology (ICT) is leading to smartification of public services (Vassilakopopolou et al., 2022). Smartification processes that are parallel to advancement in digital domain relate to evolution of digital platform generation from the so-called Web 1.0, through Web 2.0 to the currently explored Web 3.0 elaborated later in the chapter. Smart city leverages ICT to improve public services organization and delivery in municipal areas and contributes to proliferation of smartification of processes in domains of planning and management of public sector-owned tasks and obligations (Li et al., 2019).

Smart cities measurement methodologies evolved over the last decade toward direction with growing importance of digitization and connectivity aspects (OECD, 2020). The digitization of the public sector is enabled by digital infrastructure, its quality, and affordability. Such infrastructure enables the delivery of digital public services, such as government services for residents and business, transportation, healthcare, education, retail, and hospitality, and public space management and infrastructure investments planning.

Public space management in the chapter refers to the practice of management of external public spaces (Bibri et al., 2022). It is the subject of urban design, architecture, and environmental studies. Recent literature puts public space management in the context of its impact on quality of life of citizens, environmental quality, and sustainability (Prayliya & Garg, 2019). It is expected from local authorities to provide proper management oriented on assurance of quality and success of public spaces. The domain of space management in the chapter refers to public space destination, functions, and performance planning and management, including public infrastructure investments (e.g., recreational areas, residential areas, transportation routes, parking zones, and repurposing of existing spaces). Recent developments in this domain prove extensive usage of digital solutions, also in the form of advanced digital platforms, as COVID-19 pandemic contributed to the proliferation of digital solutions used by LAUs in the space management domain (Omari & Egho, 2023). In brief, smartification of public space management refers to efforts oriented toward increase of efficiency of public space management through information and communication technology and increase of knowledge and know-how distribution among the stakeholders involved in the public space management domain, such as citizens and enterprises.

6.4 Public space management and metaverse

The metaverse is defined as one of the most recent approaches in discussion on leveraging virtual reality, augmented reality, and mixed reality for the advancement of collaboration of various stakeholders in virtual environments

(Wu et al., 2022). It is defined as virtual reality space, wherein users can interact with others in a graphically rich virtual environment (Han, 2022). The metaverse industry has been growing in recent years. Major factors contributing to this growth have been technological advancements in digital infrastructure, changes in the people interactions' environment, and also the context of COVID-19 pandemic. Scholars indicated segments emerging within the metaverse industry relating to specialized metaverse platforms reflecting the various needs of heterogenous users (Lee & Gu, 2022).

In the context of public space management, metaverse platforms in the form of stable online virtual environments seem to be the next step in solutions leveraged by smart cities to better interact with their stakeholders. Scholars indicate that the metaverse type of solutions are socially constructed, politically driven, economically conditioned, and historically situated. The metaverse in its intent is always on three-dimensional (3D) network of virtual spaces where users can socialize, interact, connect, learn, or work because of convergence of data-driven immersive technology. The most recent research on the role of metaverse in smart cities points out on its contribution to an emergence of a new form of urbanism through digitally powered and data-enabled urban environments (Bibri, 2022).

Metaverse platforms classified as participatory, immersive, dynamic, and multidimensional realms for connecting people and enabling them in social interactions provide their users with a sensation of almost physical presence (Voinea et al., 2022). In recent years, the metaverse has been discussed in the context of virtualization of smart cities through the lens of its potential to redefine city design activities and service provision toward increase of efficiency, accountability, and quality of public services performance (Allam et al., 2022).

Public space management in a city involves various stakeholders with different interests and information required for them about physical city objects. Distributed data sources and data integration are a common challenge in smart cities and public space management projects they run (Knezevic et al., 2022). Recently, digital replicas of physical objects and processes proliferated and under the name of digital twins also became part of public space management and infrastructural projects planning.

Digital twins comprise the data collected from multiple sources and often also include a layer of behavioral insights received from the users of digital twin model. Digital twins enable modeling and simulation of various scenarios in a safe, virtual environment. They are particularly useful for testing "what if scenarios" for short-term, mid-term, and long-term perspectives, and, as a result of testing in a virtual environment, they can lower the costs of decision-making and help to avoid the costs of future corrective measures. Digital twin models in relation to public space management are virtually true replicas of physical current states of objects or processes. In other words, they are virtualized cities which physical citizens and enterprises may have access to for testing and discussing various scenarios and possibilities. They are oriented on improvement of management processes and allocation efficiency of public space management with the involvement of human resources and infrastructure through dedicated tools enabling processes optimization. Digital twin models of cities are pre-requisite for metaverse type of solutions from the perspective of creating a virtual 3D baseline model of a city where its users can virtually collaborate (Lv et al., 2022). Digital twins in the context of public space management provide computational decision support and might contribute to sustainability planning within city planning and governance (Corrado et al., 2022).

Digital platforms such as city official Internet portals in form landing pages belong to generation of so-called Web 1.0 digital solutions, which are oriented on one-way communication by data provision and enablement of if reception by intended audience. Digital platforms enabling dialogue, discussion, and two-way conversations belong to the next generation of digital solutions called Web 2.0. Among Web 2.0 solutions, the so-called conversational digital platforms are located. Digital twin platforms enabling discussion over digital models belong to this generation of solutions. The last generation of immersive experience digital platforms, often augmented by virtual reality solutions, represents Web 3.0 generation (also known as 3D web), which metaverse belongs to. Generations of Web 1.0, 2.0, and 3.0 are distinguished between each other by the level of participations of their users. In Web 1.0, users were passive; in Web 2.0, they were encouraged to participate; in Web 3.0, they are oriented and channeled toward cooperation (Mithun & Bakar, 2020) (Figure 6.1).

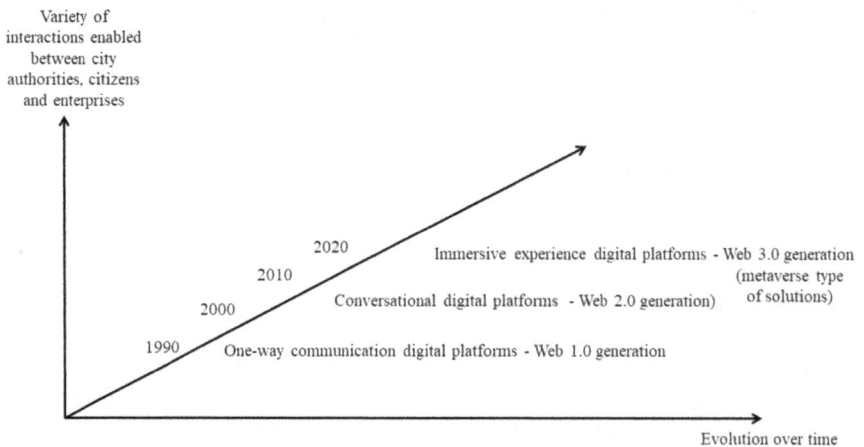

Figure 6.1 Direction of smartification process based on the types of digital platforms used for communication between city authorities, citizens, and enterprises.

Source: The author.

Based on literature review in this part of the chapter, the RQs were formulated:

RQ1: What type of solutions and practices are used by smart cities to advance public space management domain?

RQ2: What are the benefits and risks related to the usage of solutions and practices used by smart cities in the public space management domain?

RQ3: What is the direction of smartification process of public space management in relation to solutions and practices used by smart cities?

RQs are addressed through empirical research method design, empirical research results presentation, discussion, and conclusion sections in the next parts of the chapter.

6.5   Smartification of public space management research data and methods

To answer RQs, a qualitative data research was designed. The primary research method for data collection was an electronic audit of official Internet portals of 25 European cities ranked in SCI 2021. The author used purposive sampling to select the research sample from the EU highly ranked cities recognized as reference points in smart cities discourse. Electronic audit took place in September 2022 based on analyzing and coding content available in official Internet portals of the following cities: Amsterdam, Berlin, Birmingham, Dublin, Dusseldorf, Hamburg, Hannover, Kiel, Manchester, Munich, Rotterdam, Vienna, Bilbao, Bordeaux, Brussels, Copenhagen, Glasgow, Gothenburg, Helsinki, Leeds, Lille, London, Madrid, Saragossa, and Stockholm. In the electronic audit, the performance questionnaire of the following aspects relevant for RQs set was analyzed in a standardized way – smart city/smart organization projects examples, examples of technology solutions local administration unit managers, and/or citizens, forms of solutions enabling open feedback from citizens to local administration unit managers, types of cooperations and partnerships exposed in the official Internet portal (with business, education sector, other cities, etc.), and types of competitive advantages and benefits promoted by the city from general view/welcome page view in the official website.

In terms of projects and initiatives looked for, first and foremost public space and infrastructure management areas were explored. These areas referred to public space management and its improvement, also including their environmental aspects, urban planning including spatial aspects of the city. Particular attention was brought to solutions and technology enabling better co-governance, consultations, and involvement of space and infrastructure management stakeholders, such as citizens and enterprises.

A secondary research method applied were surveys performed with official representatives of cities ranked in the SCI 2021, which were oriented on deepening aspects discovered during the electronic audit of official Internet portals.

Aspects explored within the surveys referred to advantages and benefits and risks and barriers related to concrete solutions and practices used when compared them to the period when they did not exist. Surveys including direct individual interviews (IDIs) were performed in October 2022 with representatives of smart cities ranked in SCI 2021 such as Barcelona, Duesseldorf, Goteborg, Hannover, Leeds, Lisbon, Madrid, Rotterdam, and Vienna.

Primary and secondary research methods enabled to capture data leading to answering RQ1, RQ2, and RQ3. Data captured in a descriptive and explanatory manner is presented in the next subchapters of the study.

6.6 Smartification of public space management research – Landscape of solutions and practices

The audit of official Internet portals of selected European cities proved to provide arguments that currently, the portals of organizations undergo transformation from information management platforms toward conversational platforms. Examples of concrete solutions available in the Internet portals audited were identified contributing to conversational character of Internet platforms, such as rule-based chatbots and virtual assistants, content presented in the form of storytelling and data storytelling (e.g., through visualizations and infographics), and open data sets enabled to various stakeholders to make use of them for the benefit of local communities. Concrete examples of space and infrastructure management solutions were identified in cities audited and surveyed. Solutions and practices elaborated within the subchapter are structured in Table 6.1.

London (Great Britain), e.g., enabled dialogue related to early-stage drafts of strategic documents the city hall is working upon, also including public space management and spatial planning ones. Strategies and plans section in the official Internet portal of London details documents not only on housing, environment, and economy but also on health, skills, culture, food, sport, and fosters dialogue around them. City-led online communication channel in the form of proprietary social medium with engagement of local and district level

Table 6.1 Solutions and practices related to smartification of public space management are identified in the sample researched

No.	Type of solutions and practices	Cities with examples of solutions and practices identified
1	Data enablement platforms	Hamburg
2	Conversational digital platforms	London and Brussels
3	Mobile applications engaging citizens	Glasgow, Kiel, and Stockholm
4	Online participation tools engaging citizens	Copenhagen and Helsinki
5	Online suggestion boxes for citizens	Leeds and Madrid
6	Digital twin models	Goteborg and Rotterdam

Source: The author.

communities discusses plans and local development policies. Citizens and enterprises could join to discuss the initiatives. Local social medium operated in the form of an online community was called "Talk London" and concentrated on dialogue-enabling better preparation of policies and plans within the approach of "Let's make London better together."

Another city researched, Brussels (Belgium), operated its own public participation-oriented digital platform "faireBXLsamen." This local digital platform enabled reporting problems and challenges identified in the public space in the city. Platform "faireBXLsamen" indicated events and meetings dedicated to city development projects where people can discuss and consult. Projects were presented with reference to the existing or planned location on the map of Brussels and referred participants to their respective neighborhoods and communities. There was also a separate new projects submission form enabling Brussels inhabitants to ensure other initiatives.

Copenhagen (Denmark) is an example of another city, which proactively nurtured dialogue with citizens and enables reporting aspects of city development which might require intervention in the social domain, governance, and traffic through reporting to authorities' functionalities of its official Internet portal.

One of the most comprehensive sets of practices related to citizens' involvement in public space management was implemented in Hamburg (Germany) under the DigITAll initiative for connection, digitization of different applications and processes related to infrastructure projects carried out by various entities, including public authorities. The DigITAll initiative improved the provision of information for effective communication and coordination activities. The DigITAll also structures various solutions and processes in the field of infrastructure, based on the example of roads construction in Hamburg. DigITAll was part of broader ambition of "Hamburg Digital City," which expressed the direction of development ambition of Hamburg. Hamburg was distinguished through a series of projects and initiatives aiming at advanced digitization of public services. Digital city strategy of Hamburg and "Digital First" type of projects included the launch of "Urban Data Platform" enabling cooperation and creation of applications based on data enabled by public sector institutions for both public and private stakeholders. Another initiative in Hamburg, DIPAS, enabled digital participation in local development projects planning through access to digitized maps, photographs, plans, models, and geographical data. DIPAS enabled inhabitants of Hamburg to localize planned infrastructural projects and provide feedback on them. Another undertaking taking place in Hamburg, "Smart Delivery and Loading Zones," focused on transportation and logistics domain of Hamburg and optimization of urban traffic by virtual bookings, registration of planned deliveries, and reservations of delivery zones to react to increasing delivery traffic, related traffic jams, and CO_2 emission increase. The initiative introduced digital signs channeling the traffic in the city, managing no-stopping zones, and streamlining the city traffic in a possibly efficient manner. The area of transportation in Hamburg has also been embraced by digital service for ship transport registration with plans to include shipowners, notaries, and lawyers in documents exchange to decrease administrative efforts related to entering inland waterways. Hamburg was recognized with the award of "Best Cooperation Project 2022" at eGovernment competition 2022 for the unique approach toward rapid continued development of digital application and process optimization services at lower costs compared to market standards.

Helsinki (Finland) offered to its inhabitants various forms of participation and ownership which impact local development through a dedicated section at its official Internet portal, which enabled presentation of early-stage projects (also infrastructural ones), participating in public dialogue, submitting questions, suggestions, feedback, endorsements, or words of criticism. The city offered "The Participation Newsletter" and "Plan Watch" services on city planning, building, and maintaining urban infrastructure.

A similar participation, enabling solutions, was recognized in Madrid (Spain), through online participation tools "Sugerencias y Reclamaciones" (suggestions and complaints) or "Madrid participa" for continued dialogue between citizens and a city council. Leeds was a city that introduced a digital platform for residents to propose ideas for improving the quality of life in the city. A survey with city authorities revealed ambitions and plans on advancing the platform through Digital Leeds initiative oriented on the growth of participation and collaboration between citizens, communities, and organizations in the city.

In the domain of infrastructure, including environment infrastructure management, the waste management as a key domain was addressed by Kiel (Germany). Kiel realized waste management improvement support of dedicated application, which was enabling to report areas polluted with waste to trigger the intervention for waste elimination. This concrete solution fitted into broader ambition of Kiel called "Zero Waste City" which also encompassed raising environmental awareness among citizens and initiatives related to recycling, food saving, and efficient energy usage by households. The administration of Kiel fostered dialogue between authorities and citizens through digital solutions through public services and information available in digital format of mobile applications. In the field of public space management, the key area of focus is public transportation and enablement for citizens of data on parking spaces, traffic intensity, and public transportation availability.

Stockholm (Sweden) in the environment infrastructure and climate protection domain introduced in 2016 the concept of smart bins, an initiative cohesive with a brand promise of Stockholm – "The Eco-Smart City." Hundreds of waste bins in the city were fitted with solar-power mobile solutions and sensors which reported waste containers utilization space, optimized the space of waste within the containers, and saved the frequency of being emptied. In the longer perspective, such an approach can lower the costs of waste management, reduce waste emissions, and contribute to local sustainable development ambition. Stockholm realized its smart city concept through a series of environment-oriented initiatives and projects elaborated in the "Environment Program for the City of Stockholm," "Environment Program 2020–2023," and "Climate Action Plan."

The environment infrastructure management was also tackled by Glasgow (Scotland), where first of a kind construction in the form of Glasgow Smart Canal for water management through technology had been introduced. Glasgow Smart Canal sets up reference point in digital surface water drainage management system. The solution was proposed to mitigate flood risk in the city and enable environment regeneration with the usage of sensors and weather prediction technology.

The most technologically advanced solutions oriented toward advancement of public space management were identified in Goteborg (Sweden) and Rotterdam (The Netherlands). Both cities in time of IDIs performed were working on creating city and city processes digital models in the form of the so-called digital twins to simulate and forecast various initiatives in a virtual environment before their implementation. The goal of digital twin models of Goteborg and Rotterdam was to support the dynamics and quality of decision-making processes of administrative authorities.

6.7   Smartification of space management research – Benefits and risks

Investigated practices in public space management, deriving from novel technology such as digital communication platforms, digital twins, and virtual reality, can benefit several domains. Nurturing dialogue with citizens through digital channels (as in the case of Brussels or London) can empower them to participate in local development and join plans and projects. Initiatives enabling the collection of feedback from citizens (as in the example of Helsinki or Madrid) may contribute to decision-making process with perspectives of local development stakeholders and nurture trust, and cooperation between various types of stakeholders. Identified multiple initiatives enabling dialogue between city authorities and citizens indicate on the scalability of this type of solutions and their importance for participatory citizenship.

Based on concrete public space management initiatives identified (e.g., DigITall, DIPAS, Digital First, and Urban Data Platform in Hamburg), the benefits of transaction cost economics are realized in lowering costs of access to data related to infrastructural processes. Provision of data and information in digital format saves the time required for the access and search of local development stakeholders, including enterprises. Digital services and easiness of doing business can contribute to investment attractiveness of a location. Practices oriented toward environmental infrastructure and environmental protection (as in the example of Kiel, Glasgow, or Stockholm) may contribute to realization of a sustainable development paradigm.

Among the risks related to smartification of public space management, it's worth to indicate digital capabilities gap and various levels of digital literacy, which require to be continuously addressed in order to enable the utilization of

Functionalities enabled
for city authorities,
citizens, and enteprises

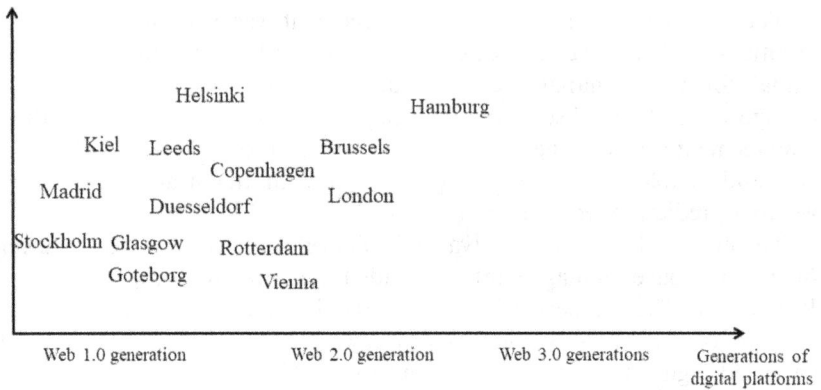

Helsinki

Hamburg

Kiel   Leeds   Brussels
                    Copenhagen
Madrid   London
              Duesseldorf

Stockholm Glasgow   Rotterdam
        Goteborg   Vienna

Web 1.0 generation   Web 2.0 generation   Web 3.0 generations   Generations of
                                                                   digital platforms

Figure 6.2 Classification of public space management solutions and practices elaborated in the chapter over digital platforms generations.

Source: The author.

novel, technology-based solutions by their intended end users. Another type of risk and limitation in the domain of public space management through novel technology solutions is their scalability depending on the budget for investments. Also change management related to adoption of new ways of working, which new solutions introduce, is a challenge to overcome when working on smartification of public space management (Figure 6.2).

6.8   Discussion, conclusions, research limitations, and new research avenues

Electronic audit of the official Internet portals of 25 smart cities ranked in SCI 2021 and nice surveys with representatives answered RQs of the chapter. Solutions advancing the public space management domain were identified, such as digital dialogue and consultation platforms, or digital simulation platforms. The best practices in the domain investigated related to empowering citizens to decision-making and impact on local development through technology, education, and digital capabilities enablement within the process of smartification (RQ1).

Smartification of public space management is not free of risks, as it requires addressing digital literacy gap and its budget is dependent on technology development and implementation perspective. The benefits of smartification of public space management are citizenship involvement, impact and co-responsibility for local development processes, and, through access to data on investments and plans, lowering the costs of information for citizens and enterprises (RQ2).

Solutions and practices related to public space management in smart cities researched reflect evolve from one-way communication, through conversational toward engagement and collaboration enabling platforms (RQ3). Deployment of digital twin models by smart cities (Goteborg and Rotterdam) strengthens theoretical aspects of the role of digital twins in smart cities elaborated in the literature (Corrado et al., 2022). Technologically advanced solutions such as DigITall, DISPAS, Digital First, Urban Data Platform (Hamburg), Plan Watch (Helsinki), Talk London (London), or FaireBXLsamen (Brussels) also prove the evolution of smart cities digital platforms into immersive, and engaging Web 3.0 and metaverse class type of solutions, outlined in the literature as solutions which may boost collaboration of various types of stakeholders (Mithun & Bakar, 2020).

To conclude, the metaverse phenomenon leveraging virtual reality, augmented reality, and mixed reality for the advancement of collaboration of various stakeholders in virtual environments appears to be a topic considered by smart cities in their development agendas due to its participation and collaboration enablement potential. The limitation of the empirical study performed is relatively small research sample (25 electronic audits and 9 IDIs), which needs to be addressed through broader research sample. Also, a purposive research sampling method was used, which limits the representative character of the findings. Empirical research boundaries refer to geographical area of Europe, which requires broader geographical area coverage to enable the conclusions to be universal and of utilitarian character. Further research avenues inspired by the literature and research performed relate to types of capabilities required and technology enablers for public space management solutions advancement, factors influencing the adoption of public space management solutions by their users (authorities, citizens, and enterprises), and appropriation of solutions and practices in the context of city location, size, and budgetary affordability. Also financing of smart city solutions in EU programing framework 2021–2027 is worth investigating in the financing context of preparations and implementations of solutions and practices explored.

References

Adamik, A.. Sikora-Fernandez, D., (2021). Smart Organizations as a Source of Competitiveness and Sustainable Development in the Age of Industry 4.0: Integration of Micro and Macro Perspective. *Energies*, 14, 1572. DOI: 10.3390/en14061572

Ajobiewe, T., (2020). Critical Assessment of Regional Development Theories. Urban Development Theories and Policies: A Critical Review and Evaluation, Ankara.

Al Omari, K., Egho, S., (2023). Redesigning the Post Covid-19 City: Management of Spaces and of Healthcare System, Distribution of Necessary Services and of Entertainment Spaces. *Civil Engineering and Architecture*, 11(1), 1–12.

Allam, Z., Sharifi, A., Bibri, S.E., Jones, D.S., Krogstie, J., (2022). The Metaverse as a Virtual Form of Smart Cities: Opportunities and Challenges for Environmental, Economic, and Social Sustainability in Urban Futures. *Smart Cities*, 5. DOI: 10.3390/smartcities5030040

Bell, S., Benatti, F., Edwards, N. R., Laney, R., Morse, D. R., Piccolo, L., Zanetti, O., (2017). Smart Cities and M3: Rapid Research, Meaningful Metrics and Co-Design. *Systemic Practice and Action Research*, 31, 27–53. DOI: 10.1007/s11213-017-9415-x

Bibri, S.E., (2022). The Social Shaping of the Metaverse as an Alternative to the Imaginaries of Data-Driven Smart Cities: A Study in Science, Technology, and Society. *Smart Cities*, 5. DOI: 10.3390/smartcities5030043

Bibri, S.E., Allam, Z., Krogstie, J., (2022). The Metaverse as a Virtual Form of Data-Driven Smart Urbanism: Platformization and Its Underlying Processes, Institutional Dimensions, and Disruptive Impacts. *Computational Urban Science*, 2, 24. DOI: 10.1007/s43762-022-00051-0

Chavarría-Barrientos, D., Molina Espinosa, J.M., Batres, R., Ramírez-Cadena, M., Molina, A., (2015). Reference Model for Smart x Sensing Manufacturing Collaborative Networks - Formalization Using Unified Modeling Language [in:] Camarinha-Matos, L.M. (eds.): *PRO-VE 2015, IFIP AICT 463*, pp. 243–254. DOI: 10.1007/978-3-319-24141-8_22

Corrado, C.R., DeLong, S.M., Holt, E.G., Hua, E.Y., Tolk, A., (2022). Combining Green Metrics and Digital Twins for Sustainability Planning and Governance of Smart Buildings and Cities. *Sustainability*, 14, 12988. DOI: 10.3390/su142012988

European Commission (2014). *EC Digital Agenda for Europe: Smart Cities*, http://eige.europa.eu/resources/digital_agenda_en.pdf, accessed on 7th of June 2022

Godlewska-Majkowska, H., Komor, A., (2019). Intelligent Organization in a Local Administrative Unit: From Theoretical Design to Reality, *European Research Studies Journal*, XXII(4), 290–307.

Godlewska-Majkowska, H., Pilewicz, T. (eds.), (2020). *Inteligentne organizacje – specyfika, rozwój i dobre praktyki przedsiębiorstw* [Smart organizations – specificity, development and the best practices of enterprises], Warsaw School of Economics Publishing House, Warsaw.

Godlewska-Majkowska, H., Pilewicz, T., Zarębski, P., (2023). *Smart Organizations in the Public Sector Sustainable Local Development in the European Union*. Routledge. ISBN 9781032209074

Han, J., (2022). An Information Ethics Framework Based on ICT Platforms. *Information*, 13, 440. DOI: 10.3390/info13090440

IMD Smart City Index, (2022). https://www.imd.org/smart-city-observatory/home/, accessed on 13 December 2022.

Jiang, M., Luo, X., Chen, C., (2018). The Factors and Growth Mechanism for Smart City: A Survey of Nine Cities of the Guangdong-Hong Kong-Macao Greater Bay Area, Proceedings of the 2018 4th International Conference on Economics, Social Science, Arts, Education and Management Engineering (ESSAEME 2018), pp. 177–183. https://doi.org/10.2991/essaeme-18.2018.34

Knezevic, M., Donaubauer, A., Moshrefzadeh, M., Kolbe, T.H., (2022). Managing urban digital twins with an extended catalog service, ISPRS Annals of the Photogrammetry, Remote Sensing and Spatial Information Sciences, Volume X-4/W3-2022 7th International Conference on Smart Data and Smart Cities (SDSC), 19–21 October 2022, Sydney, Australia.

Lee, H. J., Gu, H. H., (2022). Empirical Research on the Metaverse User Experience of Digital Natives. *Sustainability*, 14(22). DOI: 10.3390/su142214747

Li, C., Liu, X., Dai, Z., Zhao, Z., (2019). Smart City: A Shareable Framework and Its Applications in China. *Sustainability*, 11, 4346. DOI: 10.3390/su11164346

Lv, Z., Shang, W.-L., Guizani, M., (2022). Impact of Digital Twins and Metaverse on Cities: History, Current Situation, and Application Perspectives. *Applied Sciences*, 12, 12820. DOI: 10.3390/app122412820

Mithun, A., Bakar, Z.A., (2020). Empowering Information Retrieval in Semantic Web. *International Journal of Computer Network and Information Security*, 2, 41–48. Published Online April in MECS (http://www.mecs-press.org/). DOI: 10.5815/ijcnis.2020.02.05

OECD Smart City Measurement Framework, (2020). Scoping note 2nd OECD Roundtable on Smart Cities and Inclusive Growth on Smart Cities and Inclusive Growth, 3rd December 2020, https://www.oecd.org/cfe/cities/Smart-cities-measurement-framework-scoping.pdf, accessed on 13 December 2022.

Prayliya, S., Garg, P., (2019). Public Space Quality Evaluation: Prerequisite for Public Space Management. *The Journal of Public Space*, May. DOI: 10.32891/jps.v4i1.667

Sikora-Fernandez, D., (2013). The Concept of "Smart City" in Assumptions of Urban Development Policy - Polish Perspective. *Acta Universitatis Lodziensis. Folia Oeconomica*. IMD Smart City Index, https://www.imd.org/smart-city-observatory/home/, accessed on 13 December 2022.

Vassilakopoulou, P., Haug, A., Salvesen, L.M., Pappas, I., (2022). Developing human/AI interactions for chat-based customer services: lessons learned from the Norwegian government. *European Journal of Information Systems*, 32(1), 1–13. DOI: 10.1080/0960085X.2022.2096490

Visvizi, A., Lutras, M.D., (eds.), (2019). *Smart Cities: Issues and Challenges. Mapping Political, Social and Economic Risks and Threats*, Elsevier. DOI: 10.1016/C2018-0-00336-9

Voinea, G.D., Gîrbacia, F., Postelnicu, C.C., Duguleana, M., Antonya, C., Soica, A., Stănescu, R.-C., (2022). Study of Social Presence While Interacting in Metaverse with an Augmented Avatar during Autonomous Driving. *Applied Science*, 12, 11804. DOI: 10.3390/app122211804

Wu, L., Yu, R., Su, W., Ye, S., (2022). Design and Implementation of a Metaverse Platform for Traditional Culture: The Chime Bells of Marquis Yi of Zeng. *Heritage Science*, 10, 193. DOI: 10.1186/s40494-022-00828-w

Official internet portal of city of Brussels - brucity.be
Official internet portal of city of Copenhagen - international.kk.dk
Official internet portal of city of Duesseldorf - duesseldorf.de
Official internet portal of city of Glasgow - glasgow.gov.uk
Official internet portal of city of Goteborg - goteborg.se
Official internet portal of city of Hamburg - hamburg.de
Official internet portal of city of Helsinki - hel.fi
Official internet portal of city of Kiel - kiel.de
Official internet portal of city of Leeds - leeds.gov.uk
Official internet portal of city of London - london.gov.uk
Official internet portal of city of Madrid - madrid.es
Official internet portal of city of Stockholm - international.stockholm.se
Official internet portal of city of Rotterdam - rotterdam.nl
Official internet portal of city of Vienna - wien.at

7 Algorithms and geo-discrimination risk

What hazards for smart cities' development?

Ciro Clemente De Falco and Emilia Romeo

7.1 Introduction

A smart city represents a new urban framework, which integrates multiple information and communication technology (ICT) and Internet of Things solutions, to improve the quality of life of its citizens (Musa, 2018). An important driver for smart city initiatives is formed in the design, access, and use of the appropriate ICTs and facilities on a fit-for-purpose basis. It is increasingly recognized that advanced ICT use, such as the implementation of big data technology in these urban settings, may drastically change the organization of smart cities and may help leaders in determining strategies and deciding policies for the benefit of the citizen. Access to and smart use of data, in fact, are critical success factors for the urban future (Torres-Ruiz et al., 2018). Self-driving cars, traffic lights and traffic control systems, and energy and water distribution management support systems are only a few examples of the development of various forecasting, preparation, and monitoring models enhanced by deep learning and machine learning techniques for urban development in smart cities. Furthermore, the trend towards new geo-science applications, e.g., geo-design and geo-imaging, ties in with the current discussion on intelligent and smart cities, in which the use of digital technology in cyberspace lays the foundation for modern urban analysis and planning (Kourtit et al., 2017).

Essentially, a smart city emerges when urban intelligence is added to the information. Such a city has to face the challenging task to govern a complex and open spatial system as a set of (internally and externally) connected intelligent subsystems (Kashef et al., 2021). Smart city services, smart applications, and smart devices form, in fact, an ecosystem of tools and artefacts that challenge, and at times even disrupt, conventions, norms, and rites of behaviour, thus prompting diverse behavioural changes at the level of the individual, the group, and the society at large. In this view, a smart city must integrate a multidimensional perspective to manage in an intelligent and proactive way to query the complex human–technology (specifically data) relationship (Kashef et al., 2021).

DOI: 10.1201/9781003415930-10

Cities seek to harness data to rationalize and automate the operation of public services (Kamolov & Aleksandrov, 2023) and infrastructure, such as health services, public safety, criminal justice, education, transportation, and energy (Chourabi et al., 2012).

The development and implementation of algorithms to manage this kind of data is part of a move towards data-driven decision making and must be understood in the context of the "smart city" agenda. In fact, algorithms play a fundamental role in the development and operation of smart cities, enabling optimized resource management and the automation of various processes. First and foremost, since smart cities are equipped with a vast network of sensors, devices, and intelligent infrastructure that generate a large amount of real-time information, algorithms are used for data collection and analysis. Algorithms are capable of processing and analysing this data to extract valuable insights, e.g., facilitating a better understanding of resource usage patterns and traffic flows. This knowledge can be leveraged to make informed decisions regarding resource management and improving energy efficiency, waste management, urban planning, and many other aspects.

Furthermore, algorithms are utilized for optimizing public services. For instance, intelligent routing algorithms can be implemented to optimize public transportation routes, reducing travel times and air pollution. Algorithms can also assist in traffic management by adapting traffic signal timings based on the current traffic situation, thereby contributing to the reduction of traffic congestion (Engin and Treleaven, 2019). Algorithms are also employed to enhance security in smart cities. Intelligent surveillance systems utilize facial recognition and image analysis algorithms to detect suspicious behaviours and prevent criminal activities. Additionally, algorithms can be employed to analyse data from environmental sensors, monitoring air and water quality and providing timely alerts in emergency situations (Tan et al., 2020).

Lastly, algorithms are essential for citizen–government interaction. Through digital platforms, individuals can communicate with authorities, access public services, and provide feedback. Algorithms facilitate efficient management of these platforms, enabling the collection and processing of user-generated data to improve services and tailor them to the community's needs. Thus, algorithms play a key role in the development of smart cities, enabling optimized resource management, process automation, and the implementation of intelligent services. They are essential for data analysis, optimization of public services, enhancing security, and fostering citizen engagement.

However, as artificial intelligence and big data analytics increasingly replace human decision making, questions about algorithmic ethics become more pressing (Duan et al., 2019). In fact, the topic of algorithms is not new to social scientists. In recent years, critical studies of algorithmic cultures have proliferated within science and technology studies (Diakopoulos, 2015; Ziewitz, 2016; Seaver, 2018), in particular within the "Actor–Network Theory" (ANT) promoted by Latour et al. (1979). Then, urban studies scholars have extended these conversations to smart cities and data-driven urbanism (Sokhatska & Lutsiv 2023; Kitchin, 2014a, 2014b; Del Casino, 2016; Tenney and Sieber, 2016). Although presented as objective tools, according to the ANT perspective (1970), algorithms are sociotechnical devices that incorporate the perspectives of the authors involved in their creation. Thus, mentioning "Algorithmic Risk" is important to highlight the harmful consequences – intentional or unintentional – that algorithms can generate on individuals and social groups (Brauneis & Goodman, 2018). Moreover, even regarding smart cities, algorithms generate concerns about the following aspects:

1 privacy and data security, because algorithms in smart cities require a vast amount of personal and sensitive data to function properly;
2 technological dependence, given the fact that algorithm-based smart cities can become reliant on high technological reliability and if system failures or disruptions occur, there could be a significant impact on resource and service management;
3 digital exclusion when the adoption of algorithms may lead to the exclusion of those who lack access to or are unable to use digital technologies; and
4 lack of transparency and accountability regarding the decisions made by algorithms and the logic behind them.

(Brauneis & Goodman, 2018)

Thus, since the management and governance of modern, complex, and ever-rising urban agglomerations call for focused and transparent decision-making tools (Kourtit, 2014, p. 183), it seems interesting to show how algorithmic risk arises in smart cities and more specifically in urban management, addressing the scientific debate related to a new generation of policy-aware smart city research geared towards innovation and socially inclusive economic growth for sustainability (Visvizi et al., 2018). Recognizing any algorithmic management bias is crucial to enable fair and sustainable implementation of the smart city. Given that it has been shown that algorithms can discriminate individuals and entire groups of people (Brauneis & Goodman, 2018; Jeffrey, 2018), it is interesting to understand if and how algorithms can negatively impact the development of the smart city by generating mechanisms of geo-discrimination for groups and individuals. Furthermore, algorithms can exacerbate geo-discrimination by reinforcing patterns of segregation or exclusion.

Therefore, this study addresses the following research question: *What are the possible risks associated with the uncritical implementation of algorithms in the smart cities?* The remainder of this chapter is organized as follows. After this introductory section, the following paragraphs present the conceptual background of this study. The methodology used for this analysis is described in the third paragraph. In the third section, the findings are presented, followed by discussions and conclusions.

7.2 The conceptual background

7.2.1 The ANT perspective

In the 1970s, the ANT promoted by Latour and his colleagues appeared in the science and technology studies. The ANT (Latour et al., 1979) conceives the relationship between the social, material, technological, and scientific domains innovatively and considers the role of non-human actors in social processes. Technology and science domains are related to the social domain in a complex way (Latour et al., 1979). This perspective is called the "socio-material" perspective. According to this view, technology, science, and society are co-constituted. In fact, technology and science would influence social processes and the social processes would act on technology and science transforming them, adapting them, and contaminating them through norms, practices, and cultures. However, this framework should not suggest a science and technology relativistic view; in fact, it is important to highlight its "compositionalist" character aimed at understanding the specific way in which science and technology construct knowledge (Manghi, 2018). Therefore, it is not possible to analyse technology, as well as science, in a separate manner from society, given the fact that ANT conceives the world both ontologically and epistemologically (Law, 1992) as a set of networks in which the concepts of action and intentionality are untied. Actors are not who act intentionally but are the ones that modify a state of affairs by making a difference (Latour, 2007). This definition of actor, which in the ANT approach leads to the specification of the actant, allows the inclusion of human and non-human actors in this category. The networks, which are also defined as socio-material assemblages, can be consequently composed of humans, ideas, tools, technologies, etc., each one with equal possibilities of action. Therefore, technologies, as well as sciences, social institutions, and any possible objects of sociological analysis through the ANT perspective, could be seen as a product or an effect of a heterogeneous entanglement of constantly shifting relations between human and non-human actants (Latour, 2007). Even people can be conceived as an effect generated by a series of heterogeneous and interacting materials (Law, 1992). For this reason, the ANT's aim is to investigate and theorize about the emergence of these networks, their associations, their actors, and how the networks can achieve temporary stability (Callon, 1991). This approach has been used for the first time in a scientific laboratory (Latour et al., 1979) and later, in many fields such as education, marketing, linguistics (Waldherr et al., 2019), and journalism (Primo and Zago, 2015). The recognized role of non-human actors as agents has been a very fruitful starting point for analysing some of the institutions of digital society (Lupton, 2015) such as the algorithms (Kashef et al., 2021).

7.2.2 Algorithms as socio-material object

Algorithms are often presented as objective and neutral tools unrelated to cultural, economic, political, and social mistakes. This perspective is quite dangerous because it does not consider that algorithms incorporate the point of view and the needs of their developers. In the critical algorithms studies, the "opacity" concept is used to highlight the fact that the underlying rationale in algorithms construction process is often obscure. This means that the choice that an algorithm makes (a) is not always intelligible and (b) can be biased. For example, it is interesting to think about the algorithm that Amazon used to hire candidates. This algorithm systematically discriminated the women who applied because the dataset on which the tool was trained on was incorrect (Braun et al., 2016). The choices that algorithms make can have serious consequences on individuals' life. This is the reason why some researchers talk about "algorithmic risk". Since they are used in public and private sectors, their uncritical adoption could be very problematic for the population. Algorithmic risk is defined as any intentional or unintentional harmful consequence that algorithms may have on specific individuals or social groups. Algorithmic risk represents forms of biases that can produce allocation harms and/or representation harms (Crawford, 2017). The former has an economic impact, while the latter acts on a cultural level. The allocation harm occurs when the system guided by the algorithm unequally assigns opportunities and/or resources. On the other hand, when the system reinforces stereotypes or downplays specific groups, the representation harm occurs. There is a growing literature dealing with the mistakes made by algorithms. Multiple cases of bias are documented in different sectors, such as in finance, in job employment, in social policies, and in healthcare (Kang et al., 2016). This is why it would be better not to uncritically implement an algorithm, without verifying the social impacts it could have through an algorithm audit (Aragona, 2021). Since algorithms are one of the core concepts in the smart city debate, it is interesting to understand whether and how algorithms can damage part of the city. While much of the existing research on algorithmic risks has primarily emphasized the individual-level impacts, this study shifts its focus towards the territory, recognizing the possibility of geo-discrimination risks. Geo-discrimination risk refers to the potential for biased or unequal treatment of individuals or communities based on their geographic location or characteristics within a given context. It involves the use of technology, algorithms, or policies that may result in discriminatory outcomes, perpetuating or exacerbating disparities and inequalities in various aspects of urban life, such as access to resources, services (Orwat, 2020), opportunities, or decision-making processes (Gangadharan et al., 2014). Consequently, there is a growing need to examine the risk of geo-discrimination, aiming to shed light on the mechanisms and reasons behind the potential adverse outcomes or exclusion that specific regions or communities may face as a result of insufficiently designed technological solutions or urban development strategies.

7.3   Methodology

Desk research is the examination and analysis of information obtained from research that has already been carried out (Cook et al. 2002; Faruk et al., 2022). In this study, the desk research included detailed analysis of literature review entailing journal articles, analysis of the books, international and national reports, or conference papers on the subject matters the potential "algorithmic risk" capable of transforming smart cities in places where opportunities and possibilities are not evenly distributed. Like other analytical methods in qualitative research, the document analysis requires that data be examined and interpreted to elicit meaning, gain understanding, and develop empirical knowledge (Corbin & Strauss, 2014; Rapley, 2018). Furthermore, "documents of all types can help the researcher uncover meaning, develop understanding, and discover insights relevant to the research problem" (Merriam, 1988; p. 118). In fact, through this technique, several web pages and web portals and other literature all related to the ICT's influence on the smart city's growth and development have also been analysed.

The analytical procedure entailed finding, selecting, appraising (making sense of), and synthesizing data contained in documents (Zhou & Nunes, 2016). The analysis yielded data – examples, quotations, or entire passages – that are then organized through content analysis (Labuschagne, 2003) into major themes, i.e., the positive influence of ICT in smart cities, the ability of ICT to reconfigure the spatial distribution of poor citizens, and the failures that can occur using the algorithms non-critically.

The access to the relevant literature was made from the research database Google Scholar. This provided diversified data from different sources, giving the possibility to perform the analysis of the impact of ICT reconfigures the spatial distribution of poor citizens, and to list the possible algorithms' mistakes occurred within the smart cities. Research articles were selected when at least one of the research questions was the main research topic of the study. Studies that were reported in other languages except for English or did not include any of designed research questions were excluded from this study.

7.4 Findings

The desk research showed that algorithms can influence the growth and development of a smart city in different ways, both positive and negative. The first one, e.g., regards the urbanization of the city. Following the banking crisis of the 1930s, the Federal Housing Administration (FHA) was founded in the United States. One of its tasks was to insure mortgages issued by private individuals. Accessing mortgages insured by the FHA was highly convenient for private individuals. Access to the mortgage, however, was granted to those who exceeded a specific score on a financial strength scale, and that score was attributed to a system of choices similar to algorithms (Greenwalt, 2018). Among the criteria used in the scoring were "relative economic stability" and "protection from adverse influences". These criteria related to the area of the house for which the mortgage was being sought. Adopting these criteria systematically penalized mortgages applied for homes in working-class neighbourhoods of cities, encouraging the suburbanization of US cities. In the long run, therefore, the adoption of this algorithm favoured the decline of some areas and the development of others. A similar phenomenon is replaying today in some US cities in a more technological guise and within the discursive framework of the smart city. IBM's Smarter Cities Challenge (Alizadeh, 2017) develops data-focused targeting systems that offer predictive analytics for the preparation of budgets for capital improvements and for determining which assets will decline in value (Scuotto et al., 2016). The Market Value Analysis (MVA) approach, implemented since 2001 in numerous cities throughout the United States, is a data-driven technology utilized for spatial governance and development guidance. It falls under the umbrella of algorithmically generated market value assessment, wherein the MVA serves as an exemplar. By employing the MVA, public officials and private investors are influenced to select specific neighbourhoods for investment, development, and on-going public service provision (Goldstein, 2011). Unfortunately, this preference may come at the expense of other neighbourhoods that they might choose to divest from or strategically invest in through means such as environmental amenities. These areas are often deemed financially risky. At a time of solid disinvestment in public spending in American cities today, many policymakers are turning into the systems mentioned above to decide where to allocate public and private investments and services. Although presented as objective tools capable of correctly mapping the city, these have their criticalities. These systems, for instance, are built on indicators that reproduce an economical and private-sector logic that does not take into account the other dimensions of quality of life and the public good. In fact, even if theoretically abolished, redlining practices are reintroduced through these algorithms, which, far from being neutral, condemn entire portions of cities to abandonment (Safransky, 2020). This is a clear example of how an algorithm used for urban development and resource allocation in smart cities can generate geo-discrimination. If algorithms prioritize certain areas for resource allocation or public services based on biased data, it can lead to a disproportionate distribution of benefits and resources across different parts of the city.

Moreover, through the analysis, it emerged that these are not the only ways in which algorithms can discriminate entire portions of a city. Classifying areas of the city using predictive models can also be problematic for public order management. Predictive policing algorithms, in fact, are already implemented by law enforcement agencies in European states and make it possible to identify the time and place of crimes that will be committed or the perpetrators of a crime through the analysis of a pre-existing crime series (Joh, 2019). In an analysis of the PredPol system by the Human Rights Data Analysis Group (HRDAG 2016), it was pointed out that the goal of any machine learning algorithm is to develop models from the data that is fed into the software. Therefore, when fed with police data, the algorithm learns based on this data: this means that the system does not develop models relating to offences but rather models relating to how the police record crime. Since the police record crime unevenly in a city, the models of recorded crime may differ substantially from the actual crime models. Consequently, the pattern the algorithm is likely to learn is that most crime occurs in over-represented places, which may not necessarily be the places with the highest crime levels (Brayne & Christin, 2021).

Similarly, the data based on arrests could bias the results of predictive algorithms because the police arrest more people in neighbourhoods where ethnic minorities are most present, leading the algorithms to increasingly direct police activities to those areas. Accordingly, predictive tools misdirect police patrols, and some neighbourhoods are unfairly designated crime hotspots while others are undervalued. In this context, it was suggested that training algorithms on victim reports rather than arrest data could reduce the bias of predictive systems. This is because, in theory, victim reports should be less biased because they are less influenced by police biases or feedback loops. However, a study on a predictive algorithm shows that predictive processing based on victim reports leads to biases in the results (Brayne & Christin, 2021). Comparing the system's predictions based on complaint data with actual crime data for certain areas, significant errors appeared. Thus, in an area where few crimes were reported, the tool predicted only about 20% of the actual hot spots. Conversely, in an area where many crimes were reported, the tool predicted 20% more hot spots than actually existed.

Furthermore, it is evident that victim reporting is also related to the community's trust or mistrust of the police. Thus, if you are in a community with a historically corrupt police department or with notoriously racist prejudices, this will substantially influence the crime data. In this case, the predictive system may underestimate the level of crime in a given area, thus increasing the feeling of impunity and crime itself.

So, it seems clear that several doubts have been raised concerning the accuracy of the results produced by predictive policing systems, their impact in the fight against crime, and, above all, with regard to the risk of discrimination. Some human rights associations therefore regard predictive policing as an inadmissible use of artificial intelligence (European Digital Rights EDRi, 2020).

Furthermore, from the analysis that emerged, today, apps-based on algorithms assist human actors in many everyday choices, mostly regarding the mobility in the urban space. For example, some geo-apps suggest which route to take to get home or where to go if we want something to do. The production of volunteered geography information (Goodchild, 2007) can be very useful for the management of the city. In fact, systematic tracking of Yelp restaurant reviews can inform city health inspectors about food-borne illnesses emerging from the restaurants in their jurisdictions (Glaeser et al. 2018). However, the geo-app may produce undesirable effects because of urban flows thus generated by the automated decisions made by apps and their suggestion systems. Although the representation of space is regarded as an objective practice, it encompasses cultural, political, and economic logic. For instance, the spatiality that characterizes Google maps prioritizes economic interactions (Luque-Ayala & Neves Maia, 2019). It is known that urban flows can determine the vitality of entire areas. Thus, by deciding flows, these apps can redevelop or degrade entire portions of the city or even gentrify them (Jansonn, 2019).

This brief explorative excursus in the literature highlights that the uncritical use of algorithms can lead to the discrimination of entire portions of territory, a practice defined elsewhere as geo-discrimination (Borges et al., 2021). Alongside this discrimination cited above, another one emerged and does not affect the area as a whole but occurs when individuals are discriminated against as residents of a given territory. The analysis has shown how algorithms can discriminate against entire groups, such as women or African Americans. It is known that part of the population is aware of these discriminations and try to resist them by deploying different behaviour, such as the practice of black people when applying for a job position changing their surname and all the information from which their ethnicity can be derived. These are only a few practices that aim to conceal people's ethnicity to avoid the automatic discrimination derived from the algorithm. In addition to these "discrimination variables", it is possible to add territorial discrimination. In the economic field, it is possible to distinguish two types of discrimination: geo-blocking and geo-pricing (Borges et al., 2021). The term geo-pricing refers to the way in which the pricing of a good or product varies according to the geographical location of the user. Geo-blocking refers to the technology that restricts the access to Internet content, based on the user's geographical location (Flórez Rojas, 2018). In this way, it is possible to preclude to the user, based on its geographical location, products, or services (Borges et al., 2021). For example, a hotel can hide its rooms to people that belong to a particular area, because it does not consider them trustworthy. That is what happened in Brazil. Although geo-blocking originated in the context of the analysis of services and products offered on the net, it can also be used for the cases in which the probability of obtaining a good or service, such as a loan, decreases depending on the area of residence. In these cases, the algorithm, far from being neutral, could place a sort of stigma on individuals.

7.5   Discussion and conclusion

The desk research highlighted the different applications scopes of algorithms in a smart city context to manage data in order to innovate and to create benefit for the citizen. If it is true that algorithms can promote the linear development of smart cities, their uncritical adoption in some cases can lead to non-linear growth characterized by geo-discrimination. The application of algorithms to the urban context of a smart city can provide numerous solutions in different areas, ranging from improving urban management and decision-making support, to launching new or improved services for citizens and creating new economic opportunities. Thus, AI and algorithms within smart cities can have a far-reaching impact in numerous application areas, many of which are crucial for city management and urban development and include (though not exclusively) local government, health, security, mobility, and energy (Visvizi & Lytras, 2019). Therefore, it seems to be natural to expect the implementation of algorithms in the smart city context, among other things, to foster efficiency, improve governance, and promote democratic engagement and environmental sustainability. However, regarding the research question, the findings have shown that their application in urban development is characterized by a series of risks, some of which are shared by other digital technologies. The documents analysis findings suggested a series of theoretical–conceptual results regarding the collected data.

As shown in Figure 7.1, the results obtained through this research can give birth to a conceptual framework capable of highlighting, in line with the research question, the two levels at which the risks of using algorithms can significantly impact urban space and individuals' lives generating geo-discrimination.

The first is the city level in which algorithms influence in a non-linear way the structure of the smart city, giving rise to dynamics such as suburbanization, abandonment and/or degradation of entire portions of the city, gentrification, and the definition of high-risk zones. On the other hand, the second level regards the individuals who belong to a specific urban area within the smart city and may incur phenomena such as geo-blocking, geo-pricing, or other discriminatory ones.

In this regard, the noncritical use of predictive systems influences the processes of city growth and development through penalization and/or disregarding important social factors. It is important to highlight the performance-related

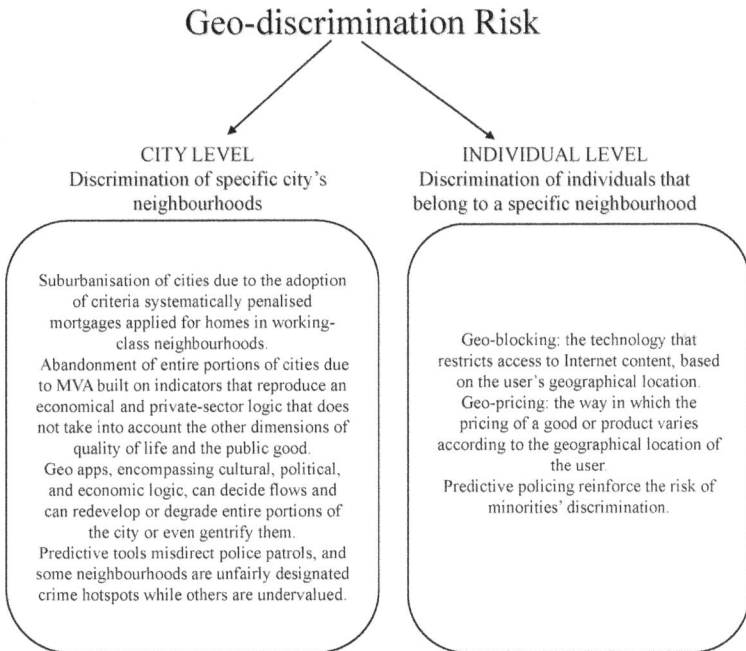

Figure 7.1 Levels of geo-discrimination risk impact.

Source: Authors.

risks that concern the so-called black box effect created by self-learning AI algorithms (Ahmad et al., 2022), which can generate or reproduce biases and lead to incorrect decisions. Eventually, other risks emerged are economic, such as the controversial displacement effect of AI. Therefore, from the study, it emerges that the complete and uncritical reliance on algorithms for the transition to smart cities can be a problem as it can lead to spatial management criticalities and impact the quality of life of individuals negatively. In particular, referring to the 11th goal set by the United Nations in the 2030 Agenda (United Nations, 2015), these criticalities undermine the fulfilment of the smart city aimed at creating benefits for citizens and the environment. Without control, smart cities can betray their own intentions and even increase inequalities through geo-discrimination. It seems clear that the relationship between smart cities and algorithms offers various great challenges and opportunities for urban development, but puts, at the same time, enormous pressure on urban areas by inducing negative externalities also, such as security issues and social discrimination and/or degradation (Visvizi et al., 2018).

From a theoretical point of view, this work sheds lights on the need for more structured and output-oriented dialogue among a variety of stakeholders on new global and socially aware polices for research on smart cities (Visvizi et al., 2018). Indeed, questions about both the politics of data creation and the definition of valid data are sparking new urban concerns and debates – over data generation, management, ownership, transparency, and access; over digital divides and the distributional consequences of "smart" technologies; over the ethics and politics of public-sector collaboration with private-data owners and managers; over the accountability, transparency, bias, and fairness of algorithmic design; and over racial bias in the technologies themselves (Diakopoulos, 2015; Ziewitz, 2016). Moreover, the work makes clear for practitioners that adopting an algorithm cannot be seen as a purely technical decision but rather a political one with political consequences. Therefore, policymakers should take into account transparency of the algorithms used, and their intelligibility should be a default criterion to be adopted. Finally, it should be emphasized that the uncritical adoption of algorithms for managing the public good would mean, in some cases, an unconscious adoption of private values within public action. This is why it would be advisable for the public administration to start investing in specialist AI departments that are able to build algorithms in compliance with public values.

References

Ahmad, K., Maabreh, M., Ghaly, M., Khan, K., Qadir, J., & Al-Fuqaha, A. (2022). Developing future human-centered smart cities: Critical analysis of smart city security, Data management, and Ethical challenges. *Computer Science Review*, 43, 100452. https://doi.org/10.1016/j.cosrev.2021.100452

Alizadeh, T. (2017). An investigation of IBM's Smarter Cites Challenge: What do participating cities want? *Cities*, 63, 70–80. https://doi.org/10.1016/j.cities.2016.12.009

Aragona, B. (2021). *Algorithm Audit: Why, What, and How?* (1st ed.). Routledge.

Borges Fortes, P. R., Martins, G. M., & Oliveira, P. F. (2021). Digital geodiscrimination: How algorithms may discriminate based on consumers' geographical location. *Droit et société*, 107, 145. https://doi.org/10.3917/drs1.107.0145

Braun, A., Zweck, A., & Holtmannspötter, D. (2016). The ambiguity of intelligent algorithms: job killer or supporting assistant. *European Journal of Futures Research*, 4(1), 1–8. https://doi.org/10.1007/s40309-016-0091-3

Brauneis, R., & Goodman, E. P. (2018). Algorithmic transparency for the smart city. *Yale Journal of Law & Technology*, 20, 103.

Brayne, S., & Christin, A. (2021). Technologies of crime prediction: The reception of algorithms in policing and criminal courts. *Social Problems*, 68(3), 608–624. https://doi.org/10.1093/socpro/spaa004

Callon, M. (1991). Techno-economic networks and irreversibility. In Law, J (Ed.), *A Sociology of Monsters: Essays on Power, Technology and Domination* (pp. 132–161). Routledge.

Chourabi, H., Nam, T., Walker, S., Gil-Garcia, J. R., Mellouli, S., Nahon, K., & Scholl, H. J. (2012, January). Understanding smart cities: An integrative framework. In *2012 45th Hawaii international conference on system sciences* (pp. 2289–2297). IEEE. https://doi.org/10.1109/HICSS.2012.615

Cooke, E., Hastings, G., & Anderson, S. (2002). *Desk Research to Examine the Influence of Marketing and Advertising by the Alcohol Industry on Young People's Alcohol Consumption: Research Prepared for the World Health Organization.* Glasgow: Centre for Social Marketing.

Corbin, J., & Strauss, A. (2014). *Basics of Qualitative Research: Techniques and Procedures for Developing Grounded Theory.* Sage Publications.

Crawford, K. (2017). The Trouble with Bias, In *31th Conference on Neural Information Processing Systems (NIPS).* Long Beach, CA, USA.

Del Casino Jr, V. J. (2016). Social geographies II: robots. *Progress in Human Geography*, 40(6), 846–855.

Diakopoulos, N. (2015). Algorithmic accountability: Journalistic investigation of computational power structures. *Digital Journalism*, 3(3), 398–415. https://doi.org/10.1080/21670811.2014.976411

Duan, Y., Edwards, J. S., & Dwivedi, Y. K. (2019). Artificial intelligence for decision making in the era of Big Data–evolution, challenges and research agenda. *International Journal of Information Management*, 48, 63–71. https://doi.org/10.1016/j.ijinfomgt.2019.01.021

EDRi. (2020). Can the EU make AI "trustworthy"? No – But they can make it just. Available at: https://edri.org/our-work/can-the-eu-make-ai-trustworthy-no-but-they-can-make-it-just/

Engin, Z., & Treleaven, P. (2019). Algorithmic government: Automating public services and supporting civil servants in using data science technologies. *The Computer Journal*, 62(3), 448–460. https://doi.org/10.1093/comjnl/bxy082

Faruk, O., Haque, N., Heuermann, A., & Al Noman, A. (2022, March). The impact of digital media and artificial intelligence on the smes in developing countries: An exploratory desk study. In *2022 8th International Conference on Information Management (ICIM)* (pp. 207–211). IEEE. https://doi.org/10.1109/ICIM56520.2022.00045

Flórez Rojas, M. L. (2018). Are online consumers protected from geo-blocking practices within the European Union?. *International Journal of Law and Information Technology*, 26(2), 119–141. https://doi.org/10.1093/ijlit/eay004

Gangadharan, S. P., Eubanks, V., & Barocas, S. (2014). Data and Discrimination: Collected Essays. *Open Technology*.

Glaeser, E. L., Kominers, S. D., Luca, M., & Naik, N. (2018). Big data and big cities: The promises and limitations of improved measures of urban life. *Economic Inquiry*, 56(1), 114–137. https://doi.org/10.1111/ecin.12364

Goldstein, I. (2011). Market value analysis: A data-based approach to understanding urban housing markets. *Putting Data to Work: Data-Driven Approaches to Strengthening Neighborhoods, 49.*

Goodchild, M. F. (2007). Citizens as sensors: the world of volunteered geography. *GeoJournal*, 69, 211–221. https://doi.org/10.1007/s10708-007-9111-y

Greenwalt (2018) The History of U.S. Housing Segregation Points to the Devastating Consequences of Algorithmic Bias – https://www.newamerica.org/pit/blog/history-us-housing-segregation-points-devastating-consequences-algorithmic-bias/

Human Rights Data Analysis Group (2016). Statisticians for human rights. Available at: https://hrdag.org/2016/

Jansson, A. (2019). The mutual shaping of geomedia and gentrification: The case of alternative tourism apps. *Communication and the Public*, 4(2), 166–181. https://doi.org/10.1177/2057047319850197

Jeffrey D. (2018). Amazon scraps secret AI recruiting tool that showed bias against women. Available at: https://www.reuters.com/article/us-amazon-com-jobs-automation-insight-idUSKCN1MK08G

Joh, E. E. (2019). Policing the smart city. *International Journal of Law in Context*, 15(2), 177–182. https://doi.org/10.1017/S1744552319000107

Kamolov, S., & Aleksandrov, N. (2023). Algorithmic modeling of public recommender systems: Insights from selected cities. *Transforming Government: People, Process and Policy*, 17(1), 72–86. https://doi.org/10.1108/TG-02-2022-0025

Kang, S., DeCelles, K., Tilcsik, A. & Jun, S. (2016). Whitened Résumés: Race and self-presentation in the labor market. *Administrative Science Quarterly*, 61(3), 469–502. https://doi.org/10.1177/0001839216639577

Kashef, M., Visvizi, A., & Troisi, O. (2021). Smart city as a smart service system: Human-computer interaction and smart city surveillance systems. *Computers in Human Behavior*, 124, 106923. https://doi.org/10.1016/j.chb.2021.106923

Kitchin, R. (2014a). The real-time city? Big data and smart urbanism. *GeoJournal*, 79(1), 1–14.

Kitchin, R. (2014b). *The Data Revolution: Big Data, Open Data, Data Infrastructures and Their Consequences.* Sage.

Kourtit, K. (2014). *Competitiveness in Urban Systems – Studies on the 'Urban Century'.* VU University.

Kourtit, K., Nijkamp, P., & Steenbruggen, J. (2017). The significance of digital data systems for smart city policy. *Socio-Economic Planning Sciences*, 58, 13–21.

Labuschagne, A. (2003). Qualitative research: Airy fairy or fundamental. *The qualitative report*, 8(1), 100–103.

Latour, B. (2007). *Reassembling the Social: An Introduction to Actor-network-theory.* Oup Oxford.

Latour, B., & Woolgar, S. (1979). *Laboratory Life: The Social Construction of Scientific Facts.* Sage Publications.

Law, J. (1992). Notes on the theory of the actor-network: Ordering, strategy, and heterogeneity. *Systems Practice*, 5(4), 379–393. https://doi.org/10.1007/BF01059830

Lupton, D. (2015). *Digital Sociology.* Routledge.

Luque-Ayala, A., & Neves Maia, F. (2019). Digital territories: Google maps as a political technique in the re-making of urban informality. *Environment and Planning D: Society and space*, 37(3), 449–467. https://doi.org/10.1177/0263775818766069

Manghi, N. (2018). Breve introduzione alla lettura di Bruno Latour. *Quaderni di Sociologia*, 77, 101–106. https://doi.org/10.4000/qds.2064

Merriam, S. B. (1988). *Case Study Research in Education: A Qualitative Approach.* Jossey-Bass.

Musa, S. (2018). Smart cities-a road map for development. *IEEE Potentials*, 37(2), 19–23.

Orwat, C. (2020). *Risks of Discrimination through the Use of Algorithms.* Federal Anti-Discrimination Agency.

Primo, A., & Zago, G. (2015). Who and what do journalism? An actor-network perspective. *Digital journalism, 3*(1), 38–52.

Rapley, T. (2018). *Doing Conversation, Discourse and Document Analysis* (Vol. 7). Sage.

Safransky, S. (2020). Geographies of algorithmic violence: Redlining the smart city. *International Journal of Urban and Regional Research, 44*(2), 200–218.

Scuotto, V., Ferraris, A., & Bresciani, S. (2016). Internet of things: Applications and challenges in smart cities: A case study of IBM smart city projects. *Business Process Management Journal, 22*(2), 357–367.

Seaver, N. (2018). What should an anthropology of algorithms do?. *Cultural anthropology, 33*(3), 375–385.

Sokhatska, O., & Lutsiv, R. (2023). What does it take to build a smart sustainable city? – Modeling an algorithm of smart cities1. In *Big Data and Decision-Making: Applications and Uses in the Public and Private Sector* (pp. 203–213). Emerald Publishing Limited. https://doi.org/10.1108/978-1-80382-551-920231013

Tan, P., Mao, K., & Zhou, S. (2020). Image target detection algorithm of smart city management cases. *IEEE Access,* 8, 163357–163364. https://doi.org/10.1109/ACCESS.2020.3021248

Tenney, M., & Sieber, R. (2016). Data-driven participation: Algorithms, cities, citizens, and corporate control. *Urban Planning (ISSN: 2183-7635), 1*(2), 101–113.

Torres-Ruiz, M., Moreno-Ibarra, M., Alhalabi, W., Quintero, R., & Guzmán, G. (2018). Towards a microscopic model for analyzing the pedestrian mobility in an urban infrastructure. *Journal of Science and Technology Policy Management, 9*(2), 170–188.

United Nations (2015). *Transforming our world: The 2030 Agenda for Sustainable Development. United Nations.*

Visvizi, A., & Lytras, M. D.. (2019). Smart cities research and debate: What is in there?. In *Smart Cities: Issues and Challenges* (pp. 1–14). Elsevier.

Visvizi, A., Lytras, M. D., Damiani, E., & Mathkour, H. (2018). Policy making for smart cities: Innovation and social inclusive economic growth for sustainability. *Journal of Science and Technology Policy Management, 9*(2), 126–133. https://doi.org/10.1108/JSTPM-07-2018-079

Waldherr, A., Geise, S., & Katzenbach, C. (2019). Because technology matters: Theorizing interdependencies in computational communication science with actor-network theory. *International Journal of Communication,* (13), 3955–3975. https://doi.org/1932-8036/20190005

Zhou, L., & Nunes, M. B. (2016). Formulating a Framework for Desktop Research in Chinese Information Systems. In J. Martins & A. Molnar (Eds.), *Handbook of Research on Innovations in Information Retrieval, Analysis, and Management* (pp. 307–325). IGI Global. https://doi.org/10.4018/978-1-4666-8833-9.ch011

Ziewitz, M. (2016). Governing algorithms: Myth, mess, and methods. *Science, Technology, & Human Values,* 41(1), 3–16. https://doi.org/10.1177/0162243915608948

8 Generative AI (GenAI) and smart cities

Efficiency, cohesion, and sustainability

Marco Moreno-Ibarra, Magdalena Saldaña-Perez, Samuel Pérez Rodríguez, and Emmanuel Juárez Carbajal

8.1 Introduction

Generative artificial intelligence (GenAI) has become an important transformative tool that has significantly improved problem-solving capabilities and optimized processes, promoted efficiency and productivity in a wide range of fields, and revolutionized society (Fui-Hoon et al., 2023; Deng and Lin, 2022; Bahrini et al., 2023). Much of the interest in GenAI is due to the fact that this technology can create multiple forms of content, including text, images, and audio. For simplicity, new user interfaces create content within a few seconds. Perhaps, the most outstanding case of a successful application of this technology is ChatGPT, which in a few weeks after its launch reached 100 million active users, representing a global phenomenon from a social and technological point of view, as well as opening the debate on its regulation and the implications of its use in different contexts (Calzada, 2023). Thus, it showed that to understand and apply this technology, it is essential to understand and apply the impact of generative AI, which requires interdisciplinary teams (Epstein and Hertzmann, 2023; Kasneci et al. 2023).

Interaction with GenAI begins with a prompt, which is an instruction or phrase given to the model to elicit speech-generated responses that can take the form of representations such as text, images, or any other form that can be processed. After an initial response, the generated content can be refined by feeding back to the system any aspects that need to be detailed or personalized. Generally, this type of technology has been implemented as a chat system, where there is an interactive collaboration between the user and the system. The forerunner of GenAI may have been a chatbot called Eliza, which was developed in the 1960s to simulate conversations with psychotherapists (Rajaraman, 2023). Following the development of Eliza, various applications have been developed to facilitate human communication with computers through natural language. It is important to note that GenAI became more prominent in 2014 with the development of a machine-learning algorithm called generative adversarial networks (GANs), which enabled the creation of images, videos, and audio that convincingly resemble those generated by real people, greatly enhancing the use of this technology (Fui-Hoon et al. 2023).

DOI: 10.1201/9781003415930-11

While this new capability has created new opportunities and challenges, it is envisioned that GenAI could significantly change the way companies operate or generate code for sophisticated applications, interpret large datasets, data analysts (Hassani and Silva), and geographic analysts (Tang and Kejriwal, 2023).

Developing transformers and large language models (LLMs) has played an important role in the growth of GenAI, and it is crucial to comment on their progress (Fui-Hoon et al. 2023). LLMs were trained on billions of pages, allowing them to answer very specific questions. In addition, the transformers can trace the connections between words stored in a large number of pages. LLMs focus on developing million-parameter models, favoring the development of GenAI models that can write text in a particular author's style or generate photorealistic images from text descriptions using tools such as Dall-E, a model that generates images from text descriptions and is useful for creating diverse visual content. These types of models integrate several artificial intelligence (AI) algorithms such as natural language processing.

It is evident that GenAI will have an even greater impact on human activities by promoting greater efficiency, facilitating the handling of large amounts of data, and helping to make faster decisions and solve problems. Similarly, users can enjoy an optimized user experience by communicating through natural language, enabling seamless human–machine interaction (Rane, 2023b), which helps them to use the products or services they have access to. These businesses bring greater convenience and innovation to improve people's lives. However, this also raises two important issues: ethical concerns about technology and the unequal distribution of income, which may affect the labor market in certain industries.

The remainder of this paper is organized as follows. Section 8.2 describes how GenAI can support the management of cities and presents some advantages of this technology, making it clear that it can be very useful. Section 8.3 presents some topics in more detail, where some applications have been developed using GenAI, or where it is envisaged that it can be useful. These include energy management and sustainability, urban planning, natural disaster management, traffic management and mobility, and geographic analyses. Section 8.4 describes considerations for the use of GenAI in urban environments, describing some details that may be useful for future developments, and Section 8.5 presents the conclusions.

8.2 Generative AI to support city management

Cities face challenges, such as population growth, environmental degradation, and inadequate transportation and infrastructure, to meet the needs of their residents. In this context, they are looking for more efficient and sustainable ways to meet their needs and address their problems (Khavarian-Garmsir, Sharifi, & Sadeghi, 2023). Available technologies and concepts such as smart cities, generative AI, geographic information systems (GISs), Internet of Things (IoT), digital twins, and Big Data play a relevant role in addressing the challenges they face (Rane, 2023b).

In terms of urban management, GenAI is attractive because of its positive impact on issues such as rapid response, improved quality of life, sustainability, and security challenges (Yan et al., 2023) (see Figure 8.1). It is well known that cities need tools that allow them to respond quickly to citizen requests, manage resources and services appropriately, address problems in a timely manner, and manage crises. This ensures an agile response to citizens' requests, satisfaction, trust in institutions, crisis prevention, and support for operational efficiency to improve their quality of life. In addition, cities provide resources and services that must adapt to the needs of their inhabitants. In this sense, this kind of environment promotes citizens' service demands, such as inquiries, complaints, and comments, and requires quick and accurate responses from government agencies, in addition to being important elements for continuous service improvement (Yan et al., 2023).

From this perspective, GenAI can understand language and generate text using various corpora to train its models. These corpora allow for the integration of knowledge about different aspects, topics, and levels of detail, including geographic information. Thus, the use of this technology to address

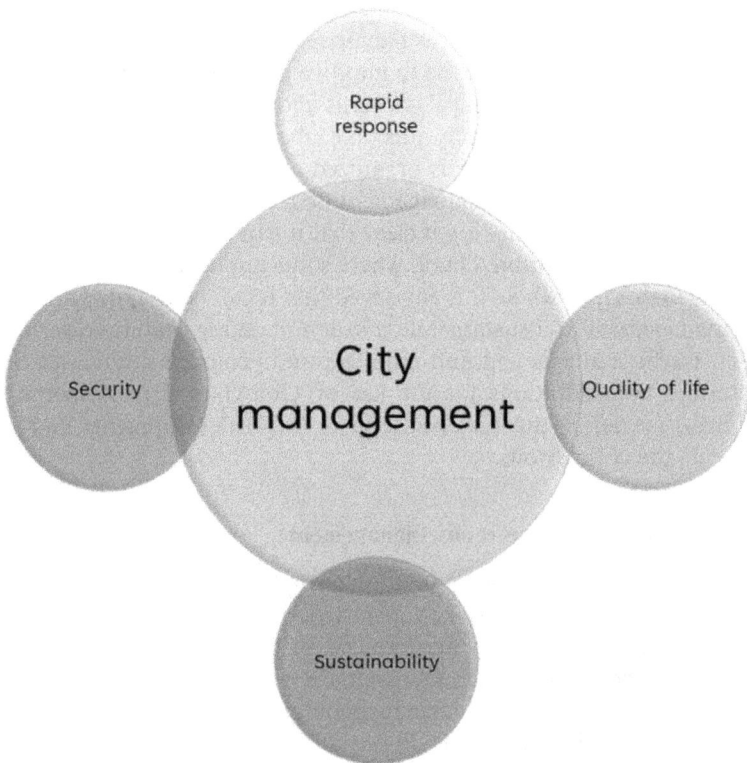

Figure 8.1 Elements of positive impact in city management.

citizen requests in the city government would significantly increase the response rate and improve the overall efficiency of the city, in addition to being a valuable tool to support decision-making and, ultimately, the definition of public policy.

It is important to note that improving citizens' quality of life is a consequence of the search for efficient urban design, personalization of services, optimization of resources, and sustainable innovation. We must not lose sight of the fact that the need for services in a city is proportional to the wealth of experiences and opportunities offered, which increases demand, and thus the dynamism of the city (Ray et al. 2024). Therefore, the incorporation of AI technologies in urban environments favors the ability to adapt to reactions and improve services through continuous iterations, where GenAI can meet these needs.

Increasing security in all aspects of life is a growing challenge for cities. Given the multiplier effect of economies of scale, cities have become important strategic spatial units driving regional and national economic growth. However, at the same time, natural disasters (Xue et al., 2023) such as earthquakes (Ahn et al., 2023) and floods, major infectious diseases such as COVID-19 (Alhasan et al., 2023), and social conflicts have a greater impact. Therefore, the use of new digital technology applications for proper management is required to provide better opportunities for the proper management of cities. GenAI provides a means for this purpose because of its analytical capabilities, which allow the analysis of large datasets (Bahrini et al., 2023).

8.3   Smart city issues through generative artificial intelligence

Smart city is a broad concept, and several definitions exist (Lytras & Visvizi, 2021). In this chapter, smart city is viewed as a city or urban area, in more general, where advanced technologies are employed to optimize resource distribution and sharing, to improve the efficiency and availability of services, to improve the quality of life of residents, and thus to foster broadly understood sustainability (Li et al., 2023, Al Fouri and Sakher, 2023; Moreno-Ibarra and Torres-Ruiz, 2019). Examples of smart city applications include traffic management, safety, energy and service optimization, and pollution monitoring (Bonomi Savignon et al., 2023; Llauradó et al., 2023; Kashef et al., 2021; Li et al., 2023). So far, the success of these types of solutions depends on how the city is monitored, how the data are integrated, what analysis is performed, and, most importantly, what results are generated to provide more factors to decision-makers. This is where GenAI can be useful for a smart city as it can integrate information from different sources, analyze it, and provide alternatives that need to be analyzed by humans to minimize the possibility of taking the wrong action. In this sense, it can address various types of problems, such as energy management, citizen security, urban planning and development, emergency management, and geographic analysis, some of which are described in the following sections (see Figure 8.2).

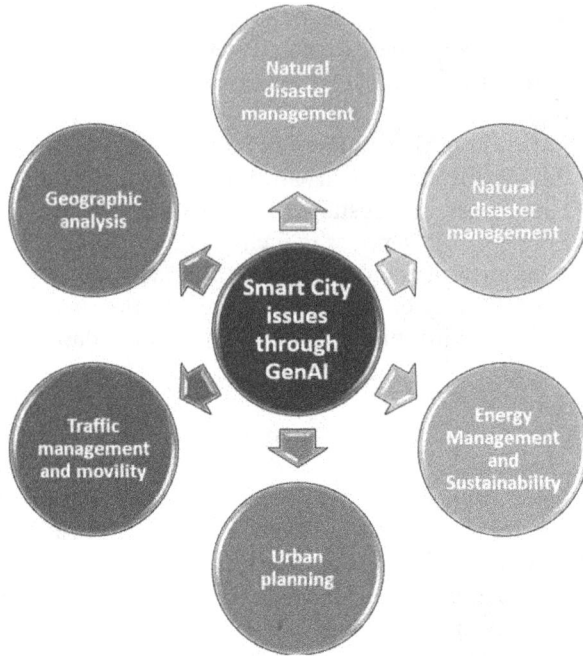

Figure 8.2 Smart city issues through generative artificial intelligence.

8.3.1　*Energy management and sustainability*

In the context of a city, GenAI applications can help with efficient energy management, which is relevant for achieving more resilient cities that promote climate change mitigation (Rane, 2023a). This technology can identify energy consumption patterns through GenAI's ability to process natural language in large amounts of textual data from documents, such as invoices, social media posts, and comments. This means that by analyzing invoices and other documents that monitor energy services, such as electricity, it is possible to identify periods and regions, as well as periods of high consumption. This can be used to promote the development of policies that encourage energy conservation and ensure a sufficient supply to meet demand, thereby reducing energy generation and achieving sustainable cities. The perception of the service can be determined by analyzing customer comments and social media posts, allowing the identification of areas for improvement or the encouragement of energy conservation.

In cities, a significant amount of energy usage is centered around buildings and industries. Therefore, taking measures to lessen consumption is key for cost-cutting and sustainability. A solution to reach these objectives would be employing GenAI, which can scrutinize data to furnish insights that can increase energy efficiency. With this cutting-edge technology, equipment data from buildings can be analyzed to locate irregularities in systems, like air conditioning, and determine causes for spikes in energy consumption or equipment malfunctioning. It can also provide suggestions to diminish energy utilization, amplify service proficiency, and, most importantly, encourage the utilization of eco-friendly sources of energy.

In general, it is possible to analyze the processes of construction, maintenance, and optimization of urban infrastructure and to identify the phases or conditions that consume more energy to promote improvements to minimize energy consumption and promote more environmentally friendly cities. Therefore, GenAI has tremendous potential for sustainable energy management because it incorporates greater analytical capabilities that provide valuable information to improve efficiency and increase sustainability.

8.3.2   *Urban planning*

With its ability to process large amounts of data and generate innovative solutions, GenAI will become a tool for analyzing environmental data, material specifications, and energy requirements to help architects and urban planners design sustainable urban spaces (Rane, 2023b; Magee and Johnson, 2023); it can as well simulate the impact of urban development projects to ensure that cities promote a better quality of life for their residents by providing them with more suitable spaces (Wang et al., 2023). Moreover, GenAI has the ability to analyze alternative scenarios and vulnerabilities. The resulting output may allow urban planners to create sustainable urban spaces that are safe and resilient to natural disasters.

GenAI has the potential to support the development of low-cost housing in a more efficient way by analyzing environment, cost, and materials, considering available space to promote sustainable development (Singh et al., 2023; Liu et al., 2023). Brozovsky et al. (2024) describe how ChatGPT can support architectural design by generating designs, which promote sustainability to reduce energy consumption by suggesting appropriate materials. In addition, they pointed out that this technology can also support the generation of project documentation, including logs, reports, and memos.

8.3.3   *Natural disaster management*

Natural disaster management considers five phases: prevention, mitigation, preparedness, response, and recovery (DOF, 2012). Xue et al. (2023) presented how the ChatGPT can support disaster management by providing real-time guidance for prevention and mitigation stages. By using GenAI, it is possible to characterize data and contextual information to create detailed contingency plans for specific scenarios. This proposal has shown how ChatGPT can integrate information from past experiences and use this knowledge for the definition of emergency plans.

It is important to note that Xue et al. (2023) comment that due to the nature of the interface and the conditions of use, it may have limitations when it comes to understanding unstructured or verbal expressions of information. This means that with this type of approach, there is a risk of not understanding subtle nuances of language, emotional expressions, or other forms of communication that may be difficult even for humans to interpret, but we believe there may be alternatives to using GenAI with some success in this area. This will undoubtedly have a positive impact on the efficiency of how cities deal with emergencies in the case of natural disasters.

8.3.4 *Traffic management and mobility*

Traffic is one of the most common problems in cities, affecting various aspects, such as the health of the inhabitants, causing stress and pollution. Several studies have proposed methods for traffic management with a sustainable approach, such as Cirianni et al. (2023) described an AI-based mobility control center (MCC) as a framework for the management, regulation, and control of urban mobility. The proposal considers the integration of IoT devices, traffic simulation and prediction software, and a Mobility as a Service (MaaS) component. MaaS consists of AI algorithms that provide personalized services with a focus on sustainability (Bharadiya, 2023). By integrating ChatGPT with MaaS, personalized recommendations can be generated based on individual preferences.

Another example would include TrafficGPT, a GenAI-based application designed to optimize traffic in the smart city (Zhang et al., 2023). By employing predominantly ChatGPT, TrafficGPT serves as an alternative to traditional modes of managing urban traffic. In other words, it supports traffic control decisions through advanced analysis of natural language interactions. In this way, it is interactive, real-time oriented, and therefore applicable incrementally (Villarreal et al., 2023).

A particular feature of GenAI is its ability to analyze heterogeneous data; therefore, it can be used to analyze reports of events that affect urban traffic. When a car accident occurs, an accident report is created that describes the event and outlines the location of the accident (Lee et al. 2018). This way of working makes it difficult to automate these reports to generate information for decision-making, as they may be unclear or ambiguous due to human intervention, where each person has a specific criterion or has more than one report for an event (Zheng et al., 2023b). GenAI enables the analysis and extraction of accident information from sets of reports that may be large, voluminous, or often incomplete (Mumtarin et al., 2023). To properly assess traffic accident reports, when there is too much missing data, data imputation is required to reach a reliable conclusion. In this sense, traditional techniques such as statistical modeling or mathematical inference are inefficient in addressing imputation in this type of data.

In this sense, using GenAI components, such as LLMs, one can take advantage of the semantic context with the images or keywords included in the reports to quickly complete the missing information as an improvement in reliability, as discussed by Zheng et al. (2023), where ChatGPT correctly identifies the cause of an accident based on social networks. Notably, GenAI can incorporate data from multiple sources to refine the extraction of details from reports. The objective of this analysis is to identify the type of accident, its location, the person involved, and its severity, which can be automated with GenAI. From what has been discussed, we have the perspective that GenAI can be a revolution in the way traffic management is performed because it has great analytical capabilities on large datasets, and the interaction is efficient; thus, efficient decision-making is possible. It is important to note that Wang et al. (2023b) describe an LLM that acts as a co-pilot during vehicle driving and can perform specific driving tasks, which may eventually be beneficial for incorporating this technology in vehicle-serving cities; also, the LLM famous model ChatGPT s considered to be used at virtual assistants implementation for intelligent vehicles, as described in Wang et al. (2023).

8.3.5 *Geographic analysis*

Recently, there has been rapid convergence between GISs and AI. Currently, the development of GenAI, particularly ChatGPT, has inspired GIS practitioners and researchers to explore how these developments can be applied to support complex decision scenarios, develop simplified geospatial modeling, and enable the natural language querying of geospatial data (Hassani & Silva 2023). The use of GenAI-based GIS can transform geospatial science by enabling natural language querying of geospatial data, automating report generation, providing suggestions for improved geospatial analysis and modeling, and providing enhanced decision support (Li and Ning, 2023).

QChatGPT is a plugin for Quantum Geographic Information System (QGIS) that integrates the ChatGPT functionality to perform tasks, such as data location, data types, and available tools. In this sense, QChatGPT can be used to determine how to use QGIS geoprocesses, which is useful for users who are just starting or requiring assistance with a specific task. It can also be used to ask questions regarding other QGIS add-ins or automated processes. Juhász et al. (2023) presented a collaborative mapping application that integrates ChatGPT and BLIP-2 with various sources of volunteered geographic information (VGI). The study shows that GenAI can help generate detailed descriptions of urban environments from photographs available on the Mapillary website and provide an alternative for enriching geographic databases of cities to integrate a greater amount of information for analysis. An interesting paper in the context of using GenAI-based GIS is presented by Mooney et al. (2023), where they evaluate the performance and ability of ChatGPT to understand geospatial concepts for a real GIS exam. ChatGPT responses were analyzed, and their understanding of the GIS principles was evaluated. The scores obtained ranged from 63.3% to 88.3%; therefore, there is room for improvement in the models.

8.4   Considerations for using GenAI in urban environments

Although the implementation of Gen(AI) represents an alternative for developing solutions to problems in cities, current developments have limitations and raised concerns in various ways. The use of GenAI raises controversies and concerns regarding model bias, privacy, model fragility, false positives, and artificial hallucinations (Dalalah and Dalalah, 2023; Zheng et al., 2023). Although these models have shown impressive performance in information extraction and missing data imputation tasks, they still exhibit moderate levels of accuracy and reliability. It is believed that GenAI models can fill in missing information with considerable accuracy using semantic context. Despite the ability of GenAI to perform tasks accurately, the generated information may still be uncertain and must be verified by humans or other models (Zhang et al., 2023).

It is important to note that in this type of development, it is difficult to understand how to adapt to new circumstances. To improve the effectiveness of this type of development in certain areas such as disaster response, it is necessary to analyze the options for accessing data in real time, considering telecommunications and other elements that may affect the arrival of timely information (Xue et al., 2023). Data quality must be considered because the usefulness and reliability of the results depend on it in GenIA implementations.

A Brazilian legislature approved the first law authored by ChatGPT. This law was presented and approved, without knowing its authorship, so when this situation became known, the debate on the use of AI in society was opened. Regulations must be established to delimit its use and to establish ethical criteria to guide its development and application (Cheng and Liu, 2023).

8.5   Conclusions

In the context of the debate on smart cities, especially in view of ensuring inclusion, safety, and resilience of urban areas, GenAI has the potential attaining substantial efficiency gains, in such domains as traffic management, disaster response, urban planning, and geographic analysis. Advances and efficiency gains in these domains may as a consequence improve the quality of life and facilitate the transition to more sustainable cities. GenAI has proven the capacity to analyze heterogeneous data, such as traffic accident reports. It is also useful for extracting information from large and incomplete data sets. Applications, such as ChatGPT, can be integrated with other, already existing applications and ICT-enhanced infrastructure components to provide personalized services. This creates several additional possibilities of how GenAI-based tools may be employed in the smart city.

Despite their benefits, there are concerns about their reliability and potential to generate misleading information that could lead to new problems (Visvizi, 2021). To reduce uncertainty and ensure the reliability of results, human review of data quality is essential. Although GenAI has remarkable capabilities, there are concerns about model bias, privacy, model fragility, and artificial hallucinations. Human oversight and attention to data quality are essential to mitigate uncertainty and ensure reliable results. As GenAI continues to evolve, its role in urban management and decision-making may expand. In summary, GenAI promises to have a positive impact on urban management, but understanding and mitigating the ethical, technical, and security challenges are essential to harnessing its benefits responsibly and effectively.

Acknowledgments

This work was supported by the Instituto Politecnico Nacional (IPN) and Consejo Nacional de Humanidades, Ciencias y Tecnologías (CONAHCYT). In addition, we would like to thank the reviewers for their valuable and constructive comments, which have helped improving the quality of this chapter.

References

Ahn, J. K., Kim, B., Ku, B., & Hwang, E. H. (2023). Virtual Scenarios of Earthquake Early Warning to Disaster Management in Smart Cities Based on Auxiliary Classifier Generative Adversarial Networks. *Sensors, 23*(22), 9209.

Al Fouri, A., & Sakher, S. (2023). Artificial intelligence on the smart city. *Remittances Review, 8*(4), 4197–4210.

Alhasan, K., Al-Tawfiq, J., Aljamaan, F., Jamal, A., Al-Eyadhy, A., Temsah, M. H., & Al-Tawfiq, J. A. (2023). Mitigating the burden of severe pediatric respiratory viruses in the post-COVID-19 era: ChatGPT insights and recommendations. *Cureus, 15*(3), 1–3.

Bahrini, A., Khamoshifar, M., Abbasimehr, H., Riggs, R. J., Esmaeili, M., Majdabadkohne, R. M., & Pasehvar, M. (2023, April). ChatGPT: Applications, opportunities, and threats. In *2023 Systems and Information Engineering Design Symposium (SIEDS)* (pp. 274–279). IEEE.

Bharadiya, J. (2023). Artificial intelligence in transportation systems a critical review. *American Journal of Computing and Engineering, 6*(1), 34–45.

Bonomi Savignon, A., Zecchinelli, R., Costumato, L., & Scalabrini, F. (2023). Automation in public sector jobs and services: A framework to analyze public digital transformation's impact in a data-constrained environment. *Transforming Government: People, Process and Policy.* https://doi.org/10.1108/TG-04-2023-0044

Brozovsky, J., Labonnote, N., & Vigren, O. (2024). Digital technologies in architecture, engineering, and construction. *Automation in Construction, 158*, 105212.

Calzada, I. (2023). Disruptive technologies for e-Diasporas: Blockchain, DAOs, data cooperatives, metaverse, and ChatGPT. *Futures, 154*, 103258.

Cheng, L., and Liu, X. (2023). From principles to practices: the intertextual interaction between AI ethical and legal discourses. *International Journal of Legal Discourse, 8*(1), 31–52.

Cirianni, F. M. M., Comi, A., & Quattrone, A. (2023). Mobility control centre and artificial intelligence for sustainable urban districts. *Information, 14*(10), 581.

D.O.F. Diario Oficial de la Federación. (2012). Ley general de protección civil. *México, DF.* https://www.gob.mx/cms/uploads/attachment/file/593503/LGPC_061120.pdf

Dalalah, D., & Dalalah, O. M. (2023). The false positives and false negatives of generative AI detection tools in education and academic research: The case of ChatGPT. *The International Journal of Management Education, 21*(2), 100822.

Deng, J., & Lin, Y. (2022). The benefits and challenges of ChatGPT: An overview. *Frontiers in Computing and Intelligent Systems, 2*(2), 81–83.

Epstein, Z., Hertzmann, A., Akten, M., Farid, H., Fjeld, J., ... & Smith, A. (2023). Art and the science of generative AI. *Science, 380*(6650), 1110–1111.

Fui-Hoon Nah, F., Zheng, R., Cai, J., Siau, K., & Chen, L. (2023). Generative AI and ChatGPT: Applications, challenges, and AI-human collaboration. *Journal of Information Technology Case and Application Research, 25*(3), 277–304.

Hassani, H., & Silva, E. S. (2023). The role of ChatGPT in data science: how ai-assisted conversational interfaces are revolutionizing the field. *Big Data and Cognitive Computing, 7*(2), 62.

Juhász, L., Mooney, P., Hochmair, H. H., & Guan, B. (2023). ChatGPT as a mapping assistant: A novel method to enrich maps with generative AI and content derived from street-level photographs. *arXiv preprint arXiv:2306.03204.*

Kashef, M., Visvizi, A., & Troisi, O. (2021). Smart city as a smart service system: Human-computer interaction and smart city surveillance systems. *Computers in Human Behavior, 124*, 106923. https://doi.org/10.1016/j.chb.2021.106923

Kasneci, E., Sessler, K., Küchemann, S., Bannert, M., Dementieva, D., Fischer, F., ... & Kasneci, G. (2023). ChatGPT for good? On opportunities and challenges of large language models for education. *Learning and Individual Differences, 103*, 102274.

Khavarian-Garmsir, A. R., Sharifi, A., & Sadeghi, A. (2023). The 15-minute city: Urban planning and design efforts toward creating sustainable neighborhoods. *Cities, 132*, 104101.

Lee, J., Abdel-Aty, M., Cai, Q., & Wang, L. (2018). Analysis of fatal traffic crash-reporting and reporting-arrival time intervals of emergency medical services. *Transportation Research Record, 2672*(32), 61–71.

Li, W., Batty, M., & Goodchild, M. F. (2023). Real-time GIS for smart cities. *International Journal of Geographical Information Science, 34*(2), 311–324.

Li, Z., & Ning, H. (2023). Autonomous GIS: The next-generation AI-powered GIS. *arXiv preprint arXiv:2305.06453.*

Liu, Y., Han, T., Ma, S., Zhang, J., Yang, Y., Tian, J., ... & Ge, B. (2023). Summary of chatgpt/gpt-4 research and perspective towards the future of large language models. *arXiv preprint arXiv:2304.01852.*

Llauradó, J.M., Pujol, F.A., Tomás, D., Visvizi. A., Pujol, M. (2023) Study of image sensors for enhanced face recognition at a distance in the Smart city context. *Scientific Reports, 13*, 14713. https://doi.org/10.1038/s41598-023-40110-y

Lytras, M. D., & Visvizi, A. (2021). Information management as a dual-purpose process in the smart city: Collecting, managing and utilizing information. *International Journal of Information Management.* https://doi.org/10.1016/j.ijinfomgt.2020.102224

Magee, D., & Johnson, K. (2023). Steamlining urban planning with AI tools. *Planning News, 49*(5), 11–12.

Mooney, P., Cui, W., Guan, B., & Juhász, L. (2023, November). Towards understanding the geospatial skills of ChatGPT: Taking a geographic information systems (GIS) exam. In *Proceedings of the 6th ACM SIGSPATIAL international workshop on AI for geographic knowledge discovery* (pp. 85–94).

Moreno-Ibarra, M., & Torres-Ruiz, M. (2019). Civic participation in smart cities: The role of social media. In A. Visvizi, & M.D. Lytras (Eds.), *Smart cities: Issues and challenges: Mapping political, social and economic risks and threats* (pp. 31–46). Elsevier. ISBN: 9780128166390. https://www.elsevier.com/books/smart-cities-issues-and-challenges/lytras/978-0-12-816639-0

Mumtarin, M., Chowdhury, M. S., & Wood, J. (2023). Large language models in analyzing crash narratives—A comparative study of ChatGPT, BARD and GPT-4. *arXiv preprint arXiv:2308.13563.*

Rajaraman, V. (2023). From ELIZA to ChatGPT: History of human-computer conversation. *Resonance, 28*(6), 889–905.

Rane, N. (2023a). Contribution of ChatGPT and other generative artificial intelligence (AI) in renewable and sustainable Energy. Available at *SSRN 4597674.*

Rane, N. (2023b). ChatGPT and similar generative artificial intelligence (AI) for smart industry: Role, challenges and opportunities for Industry 4.0, Industry 5.0 and Society 5.0. *Challenges and Opportunities for Industry*, 4. https://papers.ssrn.com/sol3/papers.cfm?abstract_id=4603234

Ray, S. S., Peddinti, P. R., Verma, R. K., Puppala, H., Kim, B., Singh, A., & Kwon, Y. N. (2024). Leveraging ChatGPT and Bard: What does it convey for water treatment/desalination and harvesting sectors?. *Desalination, 570*, 117085.

Singh, A. K., Pal, A., Kumar, P., Lin, J. J., & Hsieh, S. H. (2023). Prospects of integrating BIM and NLP for automatic construction schedule management. In *ISARC. Proceedings of the international symposium on automation and robotics in construction* (Vol. 40, pp. 238–245). IAARC Publications.

Tang, Z., & Kejriwal, M. (2023). A pilot evaluation of ChatGPT and DALL-E 2 on decision making and spatial reasoning. *arXiv preprint arXiv:2302.09068.*

Villarreal, M., Poudel, B., & Li, W. (2023). Can ChatGPT enable ITS? The case of mixed traffic control via reinforcement learning. *arXiv preprint arXiv:2306.08094.*

Visvizi, A. (2021) Artificial intelligence (AI): Explaining, querying, demystifying. In A. Visvizi, & M. Bodziany (Eds.), *Artificial intelligence and its contexts. Advanced sciences and technologies for security applications.* Springer. https://doi.org/10.1007/978-3-030-88972-2_2

Wang, D., Lu, C. T., & Fu, Y. (2023). Towards automated urban planning: When generative and chatgpt-like ai meets urban planning. *arXiv preprint arXiv:2304.03892*

Wang, S., Zhu, Y., Li, Z., Wang, Y., Li, L., & He, Z. (2023b). ChatGPT as your vehicle co-pilot: An initial attempt. *IEEE Transactions on Intelligent Vehicles*, 1–17. https://doi.org/10.1109/TIV.2023.3325300

Xue, Z., Xu, C., & Xu, X. (2023). Application of ChatGPT in natural disaster prevention and reduction. *Natural Hazards Research, 3*(3), 556–562.

Yan, Y., Li, B., Feng, J., Du, Y., Lu, Z., Huang, M., & Li, Y. (2023). Research on the impact of trends related to ChatGPT. *Procedia Computer Science, 221*, 1284–1291.

Zhang, S., Fu, D., Zhang, Z., Yu, B., & Cai, P. (2023). TrafficGPT: Viewing, processing and interacting with traffic foundation models. *arXiv preprint arXiv:2309.06719.*

Zheng, O., Abdel-Aty, M., Wang, D., Wang, C., & Ding, S. (2023b). TrafficSafetyGPT: Tuning a pre-trained large language model to a domain-specific expert in transportation safety. *arXiv preprint arXiv:2307.15311.*

Zheng, O., Abdel-Aty, M., Wang, D., Wang, Z., & Ding, S. (2023). ChatGPT is on the horizon: Could a large language model be all we need for Intelligent Transportation? *arXiv preprint arXiv:2303.05382.*

Part III

Navigating the constraints of time, space, territory, and built environment in the smart city context

9 Smart city, ICT, and older people

Developing inclusive public space and housing conditions

Ewelina Szczech-Pietkiewicz, Zofia Szweda-Lewandowska, Joanna Felczak, and Paweł Kubicki

9.1 Introduction

Information and communication technology (ICT) plays a key role in building inclusive public spaces. Examples of how ICT can be used in this context can include providing accessible information about public spaces in a variety of formats, such as audio, large print, and sign language, to make it accessible to people with different abilities. This includes information about the location, features, and services available in public spaces. For example, smart transportation systems can provide real-time information about accessible routes, and smart buildings can provide information about accessible entrances and facilities. Obviously, technology also plays a crucial role in connecting people. In this area, ICT can be used to facilitate communication and social interaction among people in public spaces (Zhang et al., 2022). For example, social media platforms and mobile apps can help people find and connect with others who share their interests and activities. From the point of view of building social capital and engaged community, ICT can be used to engage the community in the planning and management of public spaces. For example, online platforms can be used to solicit feedback and suggestions from the community and to provide information about upcoming events and activities.

When it comes to the challenge of aging society, smart cities are also a widely explored concept (European Commission, 2013; Liang et al., 2013; Smets, 2011; Stefanov et al., 2004; UNECE, 2020; Visvizi, 2023). For example, in recent years, the topic has been discussed at the United Nations Economic Commission for Europe (UNECE, 2020) and at the World Bank (Das et al., 2022), and the topic of smart cities is linked to age-friendly cities. The challenges of aging urban populations are mostly related to age-specific barriers, such as limitations for mobility, visual and hearing impairment, and chronic diseases. But smart cities can address challenges in activities that are not limited to seniors' homes, improving their quality of life in area of social participation, health care, community support, and leisure.

Against this backdrop, the objective of this chapter is to examine the possibilities that the smart city's built environment (BE) and spatial planning provide for the inclusion of the older people in urban development. The overall

DOI: 10.1201/9781003415930-13

research question is focused on finding relevant ICT and smart city solutions that will increase the quality of life of older inhabitants of cities, as well as help use the potential of the older citizens for the city development. In this area, we are focusing on the spatial planning and zoning arrangements that may improve well-being of older urban population, having in mind that aging population may be a stimulus for more coherent and linear city growth and spatial development. Moreover, the research was also motivated by questions about the relation of smart and age-friendly agendas in city policies implementation.

The value added of the discussion in this chapter lies in combining two developments: age-friendly city solutions and smart city solutions. Research in this field tends to focus on understanding the attitudes of older people toward technology (Peek et al., 2015), challenges faced by older adults in urban environments (Plouffe & Kalache, 2010; Greenhalgh et al., 2016), and developing smart city solutions that address the needs of this age group (van Hoof et al., 2018; Rocha et al., 2019). The approach in this study concerns age-friendly cities policy as a part of the whole urban development policy and answers the question of spatial management issue in smart age-friendly cities. Furthermore, it is concentrated on overstepping limitations of smart cities' growth when it comes to the demographic challenges.

The argument in this chapter is structured as follows. An overview of the challenges that demographic changes pose to urban development is presented. They include spatial planning but mostly guaranteeing competent human resources to coordinate policies directed at older people. Other issues include the position of senior-related issues in urban policies and role of bottom-up vs. top-down initiatives. The review and defined challenges allowed us to prepare a comprehensive individual in-depth interviews (IDIs) scenario, which was further developed to include the best practices in senior policy at the local level (Figure 9.1, "outcomes"). This procedure was designed to create a pool and draw conclusions on success and failure factors in the process of ICT-enhanced older citizens engagement in cities

9.2   Methods and materials

To address these issues, a series of analyses were performed. The research procedure included literature analysis concerning demographic changes and examples of living arrangements for the older people, both in the place of living and in the urban areas. Special attention was placed on the spatial planning as a factor in urban quality of life, thus answering the question of ways in which spatial aspects of city functioning have been influenced by ICT development and to bridge this study with the issue of non-linearity of urban growth. Based on literature review, an individual interview scenario was created. A series of interviews was conducted in 12 large Polish cities with three groups of stakeholders: (1) city management, (2) institutions in charge of social policy implementation, and (3) non-governmental organizations (NGOs) active in the field of older people activities. Results of the IDI's allowed us to draw

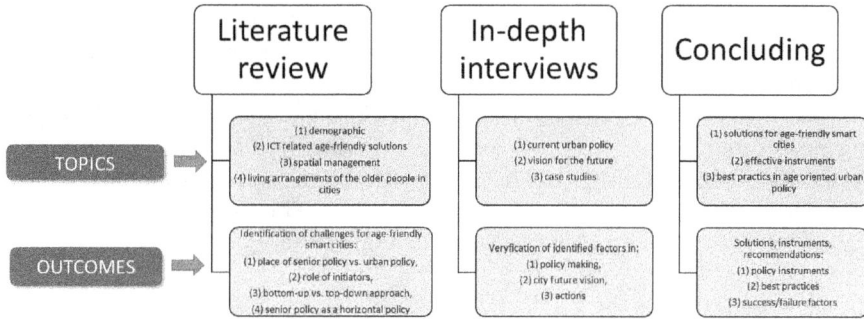

Figure 9.1 Research procedure, the logic, and the structure of the argument.

Source: Authors.

conclusions on the state of Polish cities as regards the preparation for the demographic changes. Moreover, a catalogue of good practices and ICT solutions was created to share among the cities. Finally, the implications for the further development of age-friendly smart cities were set. The research logic can be described in a graphical form, as presented in Figure 9.1, where it is represented by the "topics" row of the scheme.

Using the IDI scenario mentioned above, the research on aging urban population in Poland was conducted in the period of February–April 2022 in 12 Polish cities. The main methods used were structured IDIs and desk research. Respondents included local policymakers, experts, and NGOs. Three individual interviews were scheduled in each of the 12 cities affiliated with the Union of Polish Metropoles (UMP), one with a representative of the authorities (or a person designated by them) and two with specialists dealing with issues of aging. Typically, these were people who headed the relevant unit within the authority or were responsible for social policy. For both groups, the scenarios included a common part (a general introduction about the city development, module A on urban policy, and a conclusion with future priorities) and a separate module B. The latter, applicable to those representing the city authorities, dealt with plans for the future, and, in the case of specialists, with a selected example of action/good practice. Once the field research was completed, the interviews were analyzed, distinguishing between current activities (including good practices), the vision of development in the 2030 perspective, key actors creating urban policy, and the extent of public participation. The empirical data were supplemented by urban and regional strategies. Additionally, for the purposes of the current chapter, the analysis was expanded with quantitative data collected in the PolSenior2 survey.

The PolSenior2 study is the largest interdisciplinary study of Polish seniors. It was carried out in 2017–2020 on a sample of 5,987 Polish residents aged 60–106 (Błędowski et al., 2021). The aim of the study was to analyze the health condition of older people and their needs and to indicate recommendations that may be the basis for designing activities under public policies aimed at the seniors. The study was multidimensional and covered the medical, psychological, social, and economic aspects necessary to diagnose the situation of the older population. The study was conducted by nurses because it consisted not only of a social survey questionnaire but also of blood and urine samples from study participants, and a clock test and assessment of functioning according to the Activities of Daily Living (ADL) and Instrumental Activities of Daily Living (IADL) scales. Blood pressure measurements were also performed during three visits to the subject. This structure of the study makes it the most comprehensive study of Polish seniors.

9.3 Smart city as an age-friendly city

A smart city for an aging society is a concept that focuses on using technology and data-driven solutions to improve the lives of older individuals and support them in remaining independent and engaged in their communities. In recent years, the global population has been experiencing a significant aging trend, with the number of people over the age of 65 expected to more than double by 2050. This presents both challenges and opportunities for cities around the world.

One of the key goals of a smart city for an aging society is to use technology to improve access to services and support for older individuals. In this respect, more accessible urban services require more even spatial spread of points of access, as the main concern for older population is mobility. For example, this could involve using sensors and other technologies to monitor the health and well-being of older residents and provide alerts and support in case of emergencies. It could also involve using digital platforms to connect older residents with social and recreational activities, as well as with essential services such as transportation and healthcare.

Research suggests that the level of technology used by seniors is influenced by six major groups of factors: (1) challenges in the domain of independent living, (2) behavioral options, (3) personal thoughts on technology, (4) influence of social network, (5) influence of organizations, and (6) role of physical environment (Peek et al., 2016). Therefore, the level of ICT use in the context of aging in place is embedded in personal, social, and physical conditions of citizens. Research also shows that the attitude of older adults toward technology is usually ambivalent: there is an awareness that smart technologies may improve the quality and comfort of living but at the same time, seniors do not see how they personally need them (Lee & Coughlin, 2014; Peek et al., 2014). Older people often perceive technology as having both benefits and costs when it comes to their quality of life and independence. On the general level, technology is not seen as a means to gain or sustain independence; however, smart solutions are pointed out when it comes to experiences that improve independent living, e.g., communication, staying physically active. Generally, there is consensus in the literature that willingness to use technology in older age is influenced by pre-usage and post-usage technology-related attitudes and beliefs (Chen & Chan, 2013; Greenhalgh et al., 2013; Lee & Coughlin, 2014; Mitzner et al., 2010; Peek et al., 2014, 2016).

Barriers to the effective implementation of age-friendly smart cities initiatives can vary. The most commonly identified (Torku et al., 2020) are physical, social, technological, financial, and political. Specific subsets of barriers are interlinked and impact one another. For instance, financial measures introduced by cities during the crisis may hinder building economic security of older adults, while scarce availability of space may be in conflict in delivering age-friendly amenities as a result of policy of using resources most effectively.

Smart city solutions for older people include (Buffel et al., 2012; Das et al., 2022; European Commission, 2013; Liang et al., 2013; Smart Cities Library, 2020; UNECE, 2020):

- Smart parking systems that make it easier for seniors to find available parking spaces, reducing the need for them to walk long distances.
- Mobile apps that provide seniors with easy access to information about nearby services and amenities, such as pharmacies, grocery stores, and public transportation.
- Voice-activated assistants that can help seniors with tasks such as making phone calls, sending text messages, and managing their schedules.
- Smart lighting systems that automatically adjust the brightness and color of streetlights to make it easier for seniors to see at night.
- Sensors and other technologies that can help seniors stay connected with their families and caregivers and alert them in case of any problems.

Overall, the goal of these solutions is to help seniors live more independently and comfortably in their communities. Smart city applications can assist active aging and support cities in coping with the challenge of aging of inhabitants. The most relevant applications used in this matter concern (Rocha et al., 2019): (1) monitoring older adults, (2) supporting activities of senior citizens, (3) promotion of health and healthy lifestyle, and (4) community building.

9.4 Demographic change and the challenges cities face

The majority of countries around the world, including Poland, faces population aging. The populations of European countries are among the oldest in the world, with the share of people aged 65+ in many countries being among the highest globally. Compared to Europe as a whole, the populations of Central and Eastern European countries, including Poland, are yet "the youngest among the oldest." However, population aging will accelerate in the coming decades. For instance, according to projections, by 2060, Poland will become one of the oldest countries with the projected share of the population aged 65+ reaching 34.6% (UNDESA, 2019). This process accelerated as the post-war baby boom generation entered the old age phase. Currently, about one-fifth of Poland's population is aged 65, or more. While until 2007, the percentage of older people was higher among residents of rural areas, after 2007, the share of older people among urban residents has been dynamically increasing. In the coming decades, the aging of the population will be accompanied by the phenomenon of population decline in most cities, although this process will not be strongly differentiated spatially. Only Warsaw and Rzeszów will not record a decrease in the number of inhabitants. Other voivodeship capitals will experience a decrease in population, and the population of Poznań and Łódź in 2048 will fall below 0.5 million (GUS, 2014). The decrease in the number of inhabitants is accompanied by a change in the age structure of the inhabitants because of the acceleration not only of the aging but also of the double aging of the population. Since 1950, the proportion of people in the oldest age group (aged 85+) has risen in all European countries. The percentage of the oldest seniors, i.e., people aged 85 and over, in Poland will also increase and, in 2030, will reach 2.5% of the population. Among the post-communist countries, Poland will have the largest percentage in the oldest age group, at 5.7% in 2050.

In the next years, the double aging (defined as an increase in the share of the oldest people in the senior population) is going to accelerate (Table 9.1).

The increase in the percentage of the oldest people is important from public policy point of view because the risk of multiple diseases and limited independence increases with age, and thus the need for informal or formalized support is higher than among younger adults. Also, the living arrangements and public spaces play an important role in staying independent. The architecture barriers that are not perceived as an obstacle to live and function in a community when the person is in good health can prevent from living independently in the community as their physical strength declines.

The indicator illustrating the changes taking place in the population age structure and the challenges related to providing support is, among others, demographic dependency ratio, which shows the relationship between generations in the post-working age and generations in the working age. Currently, this ratio for the entire population of Poland is 27.5 and is below the ratio for 27 European Union countries, but the progressing aging of the population will cause its increase (Das et al., 2022). In the early 1990s, this rate was much higher in rural areas than in cities but currently, there are almost 32 people aged 65 and more in cities for every 100 people aged 15–64.

Table 9.1 Share of the oldest old (age 80, 85 years, and older) in the older population (age 65 years and older) (in %)

	Age group	
Year	*80+*	*85+*
2020	24.1	12.0
2030	26.1	11.0
2040	36.3	18.8
2050	32.8	19.7

Source: Forecast of the resident population for Poland for 2015–2050, Central Statistical Office (GUS), Warsaw 2016.

The increase in the number of people who, because of their impairment and decline in mental function, will require more or less intensive support in everyday life, will result in less and less opportunities to provide it within the household by family members. The number of one-person households inhabited mainly by older widows will increase. The migration processes that intensified after Poland joined the EU also influence the possibility of receiving support within the family network. Therefore, providing care for older persons and adjusting the living milieu is crucial. A decline in the care-giving potential of families and the rapid increase in the percentage of senior citizens make it necessary to develop community-based care designed for the older population and to adjust the public and home space for their special needs. However, the population of senior citizens is heterogeneous, which is reflected in the different reasons for providing public assistance and different paths of accessing the care system. Thus, one of the key challenges not only for the central government but also for local governments will be planning, organizing, and financing of long-term care, which is also associated with the process of deinstitutionalization of support services (Błędowski, 2021). The demographic situation in cities is the source of the threats to urban economies and their development mentioned above. Other areas of challenges include living arrangements and space planning, which can be addressed with the use of ICT. The following parts of the chapter deal with these issues.

9.5 Living arrangements of older people

The Madrid Plan, which presents the concepts of healthy and active aging, draws attention to the potential of the older population, the necessity and desirability to activate older people and involve them in the mainstream of social life (WHO, 2002). However, the concept of active aging has its roots in the idea of successful aging. The basis for successful aging is health status, family relations, relationships with friends, participation in social life, and life satisfaction (Bowling, 2007). Successful aging comes from medical science and particularly emphasizes the relation between health and life satisfaction. The three main criteria for successful aging are the ability to function independently, fitness level and the absence of chronic diseases, and an active, engaged life. Successful aging is the foundation of the concept of active aging and enables its development. In the 1970s and 1980s, the concept of active aging put emphasis on maintaining activity among the young old group at the level of activity in the middle age. Active aging popularized by the European Union in the 1990s mainly referred to creating opportunities and motivating employees to remain active in the labor market, on the one hand, while encouraging employers to hire older people, on the other. However, this concept has gradually evolved and more and more often, in addition to optimizing health opportunities enabling activity on the labor market, emphasis is placed on participation in social and family life. According to the World Health Organization, active aging is primarily the promotion of a healthy lifestyle, acting by public authorities and NGOs to adapt the common space (including public spaces) to the needs of people of different ages, including older, looking at old age and aging from a cultural, socio-economic, intergenerational, gender, and community perspective (WHO, 2002).

Adapting the space to the needs of the older people, their place of residence, and life affects the possibility of living longer in their current milieu. It allows people to grow old in a familiar space, among people they know. It also fosters activity and involvement in the life of the local community. Supporting the creation and development of institutions and organizations working for benefits of seniors by public authorities makes it possible to age in place (Ahn, 2017).

Apart from public space, the arrangement of a flat or a house is a factor determining the possibility of aging in place. With age, the time spent at home increases while housing conditions, adapted to the needs of older people with different levels of fitness, become more and more important, which affects the comfort and safety of seniors (Zrałek, 2012). The technical condition of the apartments of the older people and the provision of various amenities in both the apartments and the infrastructure of the local surroundings affect the quality of life of seniors and their independence. In the case of Polish seniors, housing has a significant impact on the independence and activity of older people. Older people often live in an old housing infrastructure, built before the economic transformation period (before 1989) and adapted to old legal requirements regarding architectural barriers and solutions facilitating the functioning of people with limited movement functions. The current, very large, aging generation of post-war baby boomers, i.e., people born in the years 1946/1947–1958/1960, live, among others, in prefabricated blocks of flats built in the previous economic system (Gronostajska, 2016).

According to the biggest research done on Polish older citizens, PolSenior2, the highest percentage of older people living in the oldest housing stock is among people over 80 years old (Błędowski et al., 2021). Only 17.9% of seniors in the 60–64 age group live in relatively new buildings, i.e., built in 1989–2000; in other age groups, this percentage is lower than 10%. At the same time, villages and small towns (up to 50,000 inhabitants) are characterized by a higher percentage of buildings built before 1944 and in the years 1945–1970, while in medium-sized and large cities, buildings from 1971 to 1988 predominate, in which facilities such as elevators, central heating, or a bathroom equipped with non-slip mats, etc., are present. A significant proportion of older people living in buildings erected in the years 1971–1978 is also related to the migration that took place in the 1970s from rural areas to cities (Błędowski et al., 2021).

One of the biggest challenges that seniors experience is the loss of mobility, problems with moving around, and climbing stairs. An important factor hindering further living in the current place, since seniors live in old houses, is the lack of elevators. One of the most frequently cited consequences of this situation is the emergence of a group of people who do not leave their apartments and are referred to in the public discussion as "prisoners of the fourth floor" from the name of the building type. The living space of these people is practically limited to the space of their own apartment. The answer to the mentioned problems is universal design. The task of universal design is to create spaces and goods aimed at ensuring their usefulness for all users, including those who, due to their age and health, are often overlooked as a group of recipients that should be considered when designing (Goldsmith, 2000).

The PolSenior2 survey also asked about the equipment of an apartment/house inhabited by an older person. Based on the answer to this question, it is possible to analyze the infrastructural deficits characterizing the living conditions of the older population and thus identify the main barriers affecting the ability of the senior to maintain independence and continue living in the current place. The presence of architectural barriers in the living place was declared by every tenth person aged 60–64, every fifth person aged 75–79, every fourth person aged 80–84, and in the age group of 85–89, the percentage of people declaring the presence of architectural barriers was almost 38%. In the oldest age group (90 years older), almost 43% of respondents reported the presence of functional barriers. The presence of architectural barriers was more often declared by city dwellers. In cities with less than 50,000 inhabitants and with a size of 50–200 thousand of residents, 17% of seniors declared the presence of architectural barriers in the building they live in, in cities with more than 200,000 inhabitants, this percentage was 20%, while only 10% of rural residents stated that there were barriers that prevent them from going out. The perception of architectural structures as barriers is also influenced by the state of health and the presence of difficulties in moving among the older people. The older without disabilities will not perceive the narrow staircase as an architectural barrier hindering their functioning, while a wheelchair user will in this case indicate its presence.

The infrastructure in the place of residence determines the possibility to live in current place and influence comfort of living of an older person. And although most seniors have access to both hot and cold running water and a flush toilet, it should be noted that the percentage of people living in places with such basic installations decreases with the age of the respondent. Significantly hindering functioning in the current place, especially in winter, is the lack of central external heating and the need to use a coal stove. The percentage of people using a coal stove increases with age. In the older age group, as many as every fourth respondent has a coal stove, which is the main source of heating the flat/house. In addition to the need to invest in the reconstruction and adaptation of housing infrastructure to the needs of older people, the use of ICT can support the independence of older people.

ICT are extensively used in gerontechnology. The use of modern technology in supporting older people is developing mainly on two levels: medical and care (Sanders et al., 2012). In the medical field, telemedicine is understood as the use of ICT to remotely monitor the vital functions and health of an older person and transfer them to a database, and thus remote exchange of information on vital parameters between an older person and a healthcare system employee (doctor, nurses, and social care worker). The area of telecare covers mainly activities aimed at the use of modern technology in order to maintain as long as possible the possibility of living in the current living environment of an older person. Telecare tools include both safety buttons and a whole system of sensors that can be installed in the apartment, home of an older person, cameras that allow viewing of rooms, sensors installed in bedding or clothes of an older person and also tablets and computers that allow contact with other people, e.g., relatives or employees of a social service or telecare center. In the field of telecare, currently in Poland mainly safety buttons are used, which are most often in the form of a wristband or as a pendant. For people with cognitive impairments, it is possible to purchase a device equipped with a GPS module, which enables localization and, in case of loss of such a person, they can be quickly found. These devices – most often associated with paying a monthly subscription – are mainly intended to quickly come to the aid of a person who has fallen and cannot get up on their own. In this case, pressing the button will cause the care/telecare center to be notified and its employee will notify the family, neighbors, or relevant services about the occurrence of a crisis situation and the need to provide assistance.

ICT in the care of the older population and in maintaining its independence will continue to develop intensively, despite two significant barriers, i.e., the cost of ICT implementation and the reluctance of older people toward ICT. With the dissemination of various solutions, the costs of using ICT will decrease over time. In case of resistance to the use of ICT-based solutions, the cohort factor, i.e., the entry old age by younger generations brought up with technologies, will contribute to the dissemination of gerontechnology. In a situation where the number of people able to provide care services is decreasing, ICT is one of the greatest opportunities to support independence in the current living place of the older people.

9.6   Aging urban population and space planning

The challenges of aging populations in cities go beyond technology-use attitudes and age-appropriate solutions. Smart cities, besides heavy use of ICT, require smart spatial planning, which is one of the priority areas in creating age-friendly cities. Since demographically adjusted policies focus on active aging so should urban planning. Therefore, active aging urban planning focuses on promoting mobility within cities such as the use of public transport and walkability, issue of safety and security of older adults, and engaging older people in community building (van Hoof et al., 2018).

When it comes to city planning with respect to aging population needs, the crucial question is whether to create communities directed specifically toward this demographic segment or rather more diversified areas. The dilemma can be traced back to pre-smart city era with American Society of Planning Officials guidelines for achieving a more balanced distribution of older population (Kaufman, 1961). General approach nowadays remains similar, with withdrawal from segregation of older people in designated urban entities or communities. Age-friendly and senior-friendly urban design embraces all generations, facilitates social interaction, and fosters sense of community (van Hoof et al., 2018).

Planning strategies for an aging population is often focused on the idea of walkability and access by pedestrians. Designing for all levels of abilities while ensuring that people can travel between places safely is vital for sustaining an autonomous and independent lifestyle of older residents. According to the study of Arup's Foresight, Research and Innovation, and Integrated City Planning, the strategies for autonomy and independence include creating walkable environments, ensuring access to transport, enabling aging-in-place, and providing wayfinding and city information (ARUP, 2019). All of these needs can be ensured by implementing a 15-minute model in city planning.

The trend related to supporting compact, small communities resulted in the popularization of the discussion on the concept of the 15-minute city. This concept assumes that residents are able to meet all their basic needs moving from home no longer than 15 minutes on foot or by bike. These needs are usually defined as work, education, shopping, recreation, and entertainment. Cities developed in this way are made up of multifunctional districts in which there are residential buildings, office buildings, schools, parks, restaurants, and public buildings next to each other. This concept is currently being implemented in practice, e.g., Paris, Barcelona (Superilles project), Melbourne (20-minute neighborhoods), and Portland (Complete Neighborhood project). Fifteen-minute cities are also recommended as one of the instruments for rebuilding urban economies after the COVID-19 pandemic by the global urban organization C40. The assumptions of such city planning can be summarized as follows:

- People in every neighborhood have easy access to goods and services, especially groceries, fresh food, and healthcare.
- Each neighborhood has a variety of housing types, sizes, and levels of affordability to diversify household types and allow more people to live closer to work.
- The inhabitants of each district can breathe clean air, free from harmful air pollution, and there are green areas that everyone can enjoy.
- More people can work close to home or remotely thanks to the presence of smaller offices, shops, and hospitality and co-working spaces.

The concept of a 15-minute city seems to be a good solution, considering the needs of the older people. On the preferences of seniors to live in housing estates specifically designed for this age group showed that the majority of adults aged 55 and older do not want to live only among their peers and away from familiar services and normal social environment. In addition, most prefer to live in small residential areas with good access to services, preferably in familiar surroundings where they can meet both older and younger people and be surrounded by family and friends (Smets, 2011).

The age-friendly city model is based on the premise of proximity. However, the introduction of this approach can add value from smart city technology incorporation. Especially, digital twins, Internet of Things (IoT), and 6G can benefit building effective 15-minute community at all stages of the process (diagnosis, implementation, use, and monitoring). The data gathered by these technologies and processed via machine learning techniques can unveil new patterns of the urban fabric (Allam et al., 2022).

Technology can play a role in supporting the goals of an age-friendly city model by making it easier for residents to access information about the services and amenities available to them within a short distance of their homes. For instance, a smartphone app could provide maps and directions to nearby parks, grocery stores, and other services and allow people to easily plan walking or biking routes to these locations. Technology can also be used to improve public transportation systems by providing real-time information about bus and train schedules, or by enabling people to use their smartphones to pay for and access public transit. Additionally, technology can be used to support car-sharing and other alternative transportation options, such as electric bikes and scooters, which can help reduce reliance on private cars.

9.7   Age-friendly cities in Poland – Challenges, instruments, and solutions: Research results

Following the results and observations from the literature review in technology use for older people in cities, they were confronted with the reality of Polish cities. The verification of research hypotheses concerned the following: urban policy, city vision for the future, and actions taken in order to increase the quality of life of the older citizens.

Respondents pointed to several challenges related to demographic changes and some of them were identified repeatedly. The challenge indicated in the interviews by the representatives of many cities is adequate support for households of older people. This challenge is difficult due to financial and legal constraints. On the one hand, the cities indicate that an appropriate alimony program is needed that would involve the family, including those absent in the country, but at the same time enable the carers to take up professional work. This is particularly important in the case of children caring for their parents with no additional help (Szczecin, Poznań, and Wrocław).

On the other hand, for families caring for the older people with at least moderate limitations in functioning, a more accessible respite policy is necessary in the form of daycare homes with meals, apartments, and temporary 24-hour support, enabling longer relief for the carer, or services provided in the home of an older person. Currently, these services are provided at a level much lower than the needs, e.g., this is indicated by Gdańsk and Katowice).

A separate group requiring help and new solutions, appearing in the statements of city experts, are older people taking care of their disabled children. Seniors without a family, whose sense of security is built based on neighborly relations and cooperation with local NGOs (Katowice, Łódź), are also a challenge. The availability of services dedicated to seniors must therefore increase dramatically. Cities try to make senior day homes or senior clubs as dense and as close to the city as, e.g., nurseries, kindergartens, or schools (Gdańsk, Wrocław, Kraków, Poznań). The availability of services for dependent people must also increase.

At the same time, cities must learn to use the resources of active seniors, and this is already happening. Interviewees participating in the research emphasized that cities are preparing for more retirement cohorts. They analyze what professions these people will be in terms of their further activity in the labor market (Warsaw), what their family situation will be (Łódź and Poznań), and prepare people just before retirement for retirement by offering them a wide range of urban activities (Wrocław and Gdańsk).

Different types of needs perceived and articulated by seniors determine the shape of "today's cities." They require the city to coordinate activities within municipal institutions as well as with the external environment and, as every city is aware, they cannot be reduced only to services for people with difficulties in everyday functioning or to the activities of a social welfare center.

The diversity of the older population influences the activities of cities and the offer addressed to this group of inhabitants. For this reason, a large group of active seniors in individual cities is interested primarily in the cultural offer, including trips out of town (Wrocław), dance parties (Rzeszów), or simply spending time with someone and talking at home or on the phone without having to participate in meetings of large groups (Bydgoszcz). Other people, on the other hand, have a need to develop, learn new things and learn, e.g., languages or acquire knowledge, develop their hobbies, or actively consume the time they have at their disposal. They want to establish new relationships, enjoy life, share their potential and knowledge by volunteering, and often want to remain active on the labor market (Gdańsk).

The needs of seniors are as diverse as those of younger people and are largely determined by personality type and habits. According to practitioners from the surveyed cities, it is not possible, and certainly very difficult, to activate people who have not been active throughout their lives and have not spent time with their loved ones, in the space of work–home–garden. Also, for these people, cities must have an appropriate offer, because they experience loneliness, loss of social bonds, lack of a sense of belonging, and it is difficult for them to build new relationships. Usually, in the event of health problems and the need to obtain care services/support in everyday functioning, this category of the older population receives help too late. These people do not know where they can get help, they do not have an extensive network of social contacts, and, by not participating in city departments aimed at seniors, they remain out of the "radar" of social workers.

This loneliness is often connected with some kind of loss: the departure of a partner, the departure of children abroad, and the loss of friendships after the end of professional activity. These experiences, as observed by the surveyed experts, are in turn associated with reduced well-being, the risk of depression, and the risk of alcohol or drug addiction. Satisfying such needs requires individual contact, "one-on-one" work, volunteers prepared to talk to the older people, the possibility of meeting older person at home, and taking care of some matter with them slowly, such as a visit to the office or library. Cities see such a need and try to meet it, e.g., by the possibility of telephone contact. This is a form that worked especially in a pandemic because it did not require direct physical contact (Wrocław and Łódź).

The diversity of needs and services for older people also sets specific tasks in the field of communication and information transfer – and in all directions. Cities try to reach seniors with information about implemented projects through various channels: through cyclically printed magazines, leaflets, and posters placed in places where seniors stay (health centers, shops, and churches), broadcast messages, as well as dedicated programs on local television or radio and also through senior clubs, various groups of older people, and quite informally through the referral network. New forms of communication are also used – e-mails and websites. Cities (Kraków and Warsaw) have particularly positive experiences in terms of increasing the IT skills of seniors, and thus their independence. In turn, programs imposed by officials often do not work, and the right design of impacts requires good communication from seniors to the office – greater public activity of seniors, participation in planning impacts, and representation of their interests in the city (Lublin).

A separate issue is awareness-raising and intergenerational interactions addressed to the entire urban community. Their goal is to change the image of old age and the older population and to increase awareness about old age. Lessons about old age are conducted with young people (Lublin and Warsaw). The aim of these activities is to counteract ageism, prevent discrimination, and present the aging process as a natural stage in life. As seniors feel discriminated against, actions are being taken in terms of equal treatment and this problem is included in documents on exclusion and stigmatization (Bydgoszcz).

Positive experiences of cities are primarily related to the activation of seniors themselves, when, with the facilitation of organizational procedures and formal support, they become local leaders and initiators of further activities and partners for local government. Negative experiences of cities are related to the fact that the offer is not adapted to the situation of the supported people, e.g., training for carers of dependent people but without providing care for dependent people for the duration of the training. The use of funds from individual subsidies is also problematic, the officials are heavily burdened with organizational matters, and the difficulty in ensuring the sustainability of the project in the uncertain financial situation of the city.

Despite the large involvement in the senior policy, cities also encounter limitations perceived as external and independent of municipal policy, the most important of which are:

- legal restrictions (e.g., inability to use school canteens to serve meals to seniors in the afternoon),
- staffing limitations (outflow of caregivers of the older persons to other countries such as Germany or Sweden), and
- financial constraints (the new Polish order, the instability of city income due to changes in the law, and the need to limit urban initiatives).

9.8 Discussion

The use of ICT in cities requires overcoming several barriers described in the literature and emerging in the research. These are financial barriers – cities burdened with regular expenses and shrinking revenues limit investment expenditures. These conclusions are confirmed by studies conducted by researchers in 287 Polish cities indicating that "the key barrier to the development of smart cities in Poland is the unsatisfactory level of residents' wealth and the difficult financial situation of cities" (Jonek-Kowalska & Wolniak 2021). There are also organizational barriers resulting from the need to use the same space by people of different ages and with different needs. Another barrier may be the ambivalent attitude of the older people to ICT described in the literature, although not necessarily related to an aversion to ICT, but an expectation of more support and a more user-friendly interface (Guner & Acarturk, 2018). Finally, great diversity in the needs, habits, and activities of seniors should be taken into consideration.

The role of cities in implementing solutions aimed at seniors and based on information technology is crucial and runs in two directions. On the one hand, cities must pursue a policy that prioritizes preparing the city for an aging population and using information technology for this purpose. On the other hand, cities must prepare their inhabitants to use modern technology in everyday life. This preparation requires education and a support system; therefore, not only the implementation is needed but also the maintenance of new solutions, which increases the expenses and the emphasis put on the effectiveness of such investments. It is also worth pointing out that seniors are flexible. The experience of the pandemic has shown that many needs of older people can be met by remote contact (cf. Choi & Lee, 2021; Silvennoinen & Heikkinen, 2023). It can therefore be concluded that as a result of this experience, the older population is now better prepared for new ICT solutions. On the other hand, and this was emphasized by interviewees in the research, there is a growing gap between those who have been able to adapt to the new situation and, for instance, maintain social contacts online and those who have completely "disappeared" from the sight and were excluded from support activities.

Social consultations and the involvement of seniors in the process of designing solutions are necessary to ensure the effective use of ICT. The experience of the surveyed cities shows that initiatives organized in response to the declared needs of the older population always find their recipients, and initiatives "imposed" by institutions are sometimes rejected as people do not participate in them. Seniors can indicate specific needs and expected solutions and to organize a group that will benefit from this solution. However, this process requires time and communication with formal and informal representatives of the older people, similarly, to designing services and organizing activities for older residents of the city.

Involving seniors and their representatives in the design of solutions makes it possible to map the diversity of needs of independent and dependent older people and to consider the diagnosis of the needs of their families and carers. Research shows that the needs of people in specific cities vary depending on the urban fabric and specific challenges (technical, communication, and housing) faced by seniors in a given space. Cities implement aging policy and achieve the same goals in different ways, and it will probably be similar in the use of information technology. Carrying out consultations in cities on the use of ICT to meet the needs of older people can identify specific ways of using ICT tailored to a given city. Certainly, each city must go through the ICT planning process independently and additionally together with its senior citizens, with the support of NGOs and the private sector.

The benefits of designing ICT solutions in specific cities include individualization that corresponds to the reality of seniors in the city – each of them uses what they need now, and these needs can change over time. The efficiency of using space by people of different ages can therefore increase from the collective perspective, thanks to solutions addressed to all residents and from the individual perspective, i.e., older people, who gain greater independence and the ability to choose services for themselves. The benefits also include a better adjustment of services to a diverse group of seniors and the ability to flexibly meet very specific needs, e.g., providing information, ensuring a sense of security, or satisfying the need for contact.

9.9   Recommendations

Presented research was conducted to draw conclusions on the level of preparation of Polish cities to the challenges of aging society. It was focused on the possibilities that ICT and smart city models offer both to the policymakers and the older people living in cities themselves. Therefore, it was possible to analyze both social and economic implications of demographic processes and observe the role of smartification and digitalization in responding these challenges. Using the conclusions on the solutions for age-friendly smart cities, effective instruments for the older citizens, and best practices in age oriented urban policy, enabled the definition of recommendations for policymaking of smart cities for older inhabitants.

The process of developing and implementing the vision of age-friendly cities by 2030 depends on several success factors (ESPON, 2019). These include:

- adjusting the services provided to the needs of the target group, which includes, i.e., using the language of the target group, adapting activities to their culture, and using non-digital information and communication channels;
- active involvement of older people in the design and implementation of policies and measures;
- close cooperation between the stakeholders involved, with the city's mediating role facilitating contacts between all stakeholders, such as social workers, health professionals, and activists;
- a positive and informal approach to older people, excluding patronizing treatment and paying attention to "problems," using the mechanisms of positive psychology;
- improving the quality of intergenerational relations.

The previously identified challenges that cities will face in the coming years and the needs indicated by the cities themselves allow for a number of recommendations for creating city development strategies, taking into account the issue of demographic changes (ESPON, 2019; OECD, 2015). They can be grouped in several thematic bundles, and many of them can be merged with smart applications solutions.

The first recommendation area concerns developing ecosystems that enable cooperation between institutional and non-institutional partners. In this process, cities should join forces with diverse interest groups, civil society organizations, social workers, activists, health organizations, and enterprises (e.g., recreational and sports facilities, restaurants), and, in such teams, jointly develop a long-term and comprehensive strategy. Such cooperation is also facilitated by network organization of work.

Creating smart cities responsive to seniors' need also requires increasing the participation of older people in social activities. This includes keeping older people active in the labor market which is crucial not only for the economy (by avoiding labor shortages) but also for older people (by maintaining their active role in society). Increasing the social participation of the older people improves their quality of life and minimizes the risk of social isolation. The activities of cities in this area may include the promotion of voluntary work, keeping the older population in employment, offering training and re-skilling, and promoting entrepreneurship. ICT solutions described in this chapter can be particularly helpful in reaching this goal.

Raising awareness of aging issues is one of the recommendations that can benefit from the ICT use. It also increases understanding of the process among residents and provides greater legitimacy. Furthermore, it helps to change the negative perception of the aging process and promotes understanding of the needs of older people among other age groups.

Considering the needs of people of different ages in spatial development plans is a recommendation that can prove to be particularly challenging. However, this can be achieved with the smart city model solutions by transforming cities into compact centers and ensuring the transport accessibility of public services and jobs. Detailed strategies in this area include, e.g., increasing the possibilities of walking around the city, integration of the economic policy of the city and its spatial development, and decentralization of city services.

Finally, creating an age-friendly smart city requires improving the accessibility of housing resources. Strategies for increasing housing accessibility can improve the quality of life for all generations. Older people can particularly benefit from such strategies as they are at increased risk of poverty and isolation and suffer from reduced mobility, which makes it difficult for them to access services and employment. Solutions from the area of digitalization and smartification are especially helpful in creating living arrangements for older people, as described in this chapter. Housing accessibility should therefore include economic accessibility (understood as a combination of the costs of residence and transport), accessibility of services in the place of residence, and accessibility of employment and transport (private and public).

References

Age-Friendly Smart Cities. (2020, May 8). *Smart Cities Library (TM)*. https://www.smartcitieslibrary.com/growing-older-in-the-city-age-friendly-smart-cities/

Ahn, M. (2017). Introduction to special issue: aging in place. *Housing and Society*, 44(1–2), pp.1–3. https://doi.org/10.1080/08882746.2017.1398450

Allam, Z., Bibri, S.E., Jones, D.S., Chabaud, D. and Moreno, C. (2022). Unpacking the '15-minute city' via 6G, IoT, and digital twins: Towards a new narrative for increasing urban efficiency, resilience, and sustainability. *Sensors*, 22(4), p. 1369. https://doi.org/10.3390/s22041369

ARUP. (2019). *Cities alive. Designing for ageing communities.*

Błędowski, P., (ed.), (2021). Deinstytucjonalizacja opieki długoterminowej w Polsce cele i wyzwania: raport. Koalicja "Na pomoc niesamodzielnym" - Związek Stowarzyszeń, Warszawa.

Błędowski, P., Grodzicki, T., Mossakowska, M., Zdrojewski, T., Instytut Gospodarstwa Społecznego, S.G.H. w., Katedra i Klinika Chorób Wewnętrznych i Chemioterapii Onkologicznej, Ś.U.M. w K., Projekt Strategiczny Starzenie i Długowieczność, M.I.B.M. i K. w. and Zakład Prewencji i Dydaktyki, G.U.M. (2021). *PolSenior2. Badanie poszczególnych obszarów stanu zdrowia osób starszych, w tym jakości życia związanej ze zdrowiem.* [online] depot.ceon.pl. Gdański Uniwersytet Medyczny. Available at: https://depot.ceon.pl/handle/123456789/21118 [Accessed 29 Jun. 2023].

Bowling A. (2007). Aspiration for older age in the 21st century: what is successful aging?, *International Journal of Aging and Human Development*, 64(3), pp. 236–297.

Buffel, T., Phillipson, C. and Scharf, T. (2012). Ageing in urban environments: Developing 'age-friendly' cities. *Critical Social Policy*, 32(4), pp. 597–617. https://doi.org/10.1177/0261018311430457

Chen, K. and Chan, A. (2013). Use or non-use of gerontechnology—A qualitative study. *International Journal of Environmental Research and Public Health*, 10(10), pp. 4645–4666. https://doi.org/10.3390/ijerph10104645

Choi, H.K. and Lee, S.H. (2021). Trends and effectiveness of ICT interventions for the elderly to reduce loneliness: A systematic review. *Healthcare*, 9(3), p. 293. https://doi.org/10.3390/healthcare9030293

Das, Maitreyi Bordia, Arai, Yuko, Chapman, Terri B., Jain, Vibhu. (2022) Silver Hues: Building Age-Ready Cities." World Bank, Washington, DC. License: Creative Commons Attribution CC BY 3.0 IGOEurostat, 2019. Population projections at national level (2019-2100). Demographic balances and indicators by type of projection and NUTS 3 region

ESPON (2019). *ACPA -Adapting European cities to population ageing: Policy challenges and best practices final report.* [online] https://www.espon.eu/sites/default/files/attachments/1.%20ACPA%20Main%20report.pdf

European Commission (2013). *The European innovation partnership on active and healthy ageing (EIP on AHA) the European innovation partnership on active and healthy ageing (EIP on AHA) innovation for age friendly buildings, cities and environments A European Innovation Partnership on Active and Healthy Ageing priority the challenge of an ageing population in European cities what are age-friendly buildings, cities, communities and environments?* [online] Available at: https://futurium.ec.europa.eu/sites/default/files/2021-10/D4%20infographic.pdf [Accessed 28 Jun. 2023].

Goldsmith, S., (2000). *Universal design. A manual of practical guidance for architects.* Oxford: Architectural Press.

Greenhalgh, T., Shaw, S., Wherton, J., Hughes, G., Lynch, J., A'Court, C., Hinder, S., Fahy, N., Byrne, E., Finlayson, A., Sorell, T., Procter, R. and Stones, R. (2016). SCALS: a fourth-generation study of assisted living technologies in their organisational, social, political and policy context. *BMJ Open*, 6(2), p. e010208. https://doi.org/10.1136/bmjopen-2015-010208

Greenhalgh, T., Wherton, J., Sugarhood, P., Hinder, S., Procter, R. and Stones, R. (2013). What matters to older people with assisted living needs? A phenomenological analysis of the use and non-use of telehealth and telecare. *Social Science & Medicine*, 93, pp. 86–94. https://doi.org/10.1016/j.socscimed.2013.05.036

Gronostajska, B.E., (2016). *Kształtowanie środowiska zamieszkania dla seniorów.* Wrocław: Oficyna Wydawnicza PW.

Guner, H. and Acarturk, C. (2018). The use and acceptance of ICT by senior citizens: a comparison of technology acceptance model (TAM) for elderly and young adults. *Universal Access in the Information Society.* https://doi.org/10.1007/s10209-018-0642-4

GUS. (2014). Prognoza dla powiatów i miast na prawie powiatu oraz podregionów na lata 2014-2050 (opracowana w 2014 r.), Warszawa 2014.

Jonek-Kowalska, I. and Wolniak, R. (2021). Economic opportunities for creating smart cities in Poland. Does wealth matter? *Cities*, 114, p. 103222. https://doi.org/10.1016/j.cities.2021.103222

Kaufman, Jerome L. (1961). *Planning and an aging population.* 148, American Society of Planning Officials.

Lee, C. and Coughlin, J.F. (2014). Perspective: Older adults' adoption of technology: An integrated approach to identifying determinants and barriers. *Journal of Product Innovation Management*, 32(5), pp. 747–759. https://doi.org/10.1111/jpim.12176

Liang, H., Liang, Y., & Shi, J. (2013). *Building a Smart, Age-Friendly Community.* 12.

Mitzner, T.L., Boron, J.B., Fausset, C.B., Adams, A.E., Charness, N., Czaja, S.J., Dijkstra, K., Fisk, A.D., Rogers, W.A. and Sharit, J. (2010). Older adults talk technology: Technology usage and attitudes. *Computers in Human Behavior,* 26(6), pp. 1710–1721. https://doi.org/10.1016/j.chb.2010.06.020

OECD. (2015). *Ageing in Cities.* OECD. https://doi.org/10.1787/9789264231160-en

Peek, S.T.M., Luijkx, K.G., Rijnaard, M.D., Nieboer, M.E., van der Voort, C.S., Aarts, S., van Hoof, J., Vrijhoef, H.J.M. and Wouters, E.J.M. (2015). Older adults' reasons for using technology while aging in place. *Gerontology,* 62(2), pp. 226–237. https://doi.org/10.1159/000430949

Peek, S. T. M., Wouters, E. J., Luijkx, K. G., & Vrijhoef, H. J. (2016). What it Takes to Successfully Implement Technology for Aging in Place: Focus Groups With Stakeholders. *Journal of Medical Internet Research,* 18(5), e98. https://doi.org/10.2196/jmir.5253

Peek, S.T.M., Wouters, E.J.M., van Hoof, J., Luijkx, K.G., Boeije, H.R. and Vrijhoef, H.J.M. (2014). Factors influencing acceptance of technology for aging in place: A systematic review. *International Journal of Medical Informatics,* [online] 83(4), pp. 235–248. https://doi.org/10.1016/j.ijmedinf.2014.01.004

Plouffe, L. and Kalache, A. (2010). Towards global age-friendly cities: Determining urban features that promote active aging. *Journal of Urban Health,* 87(5), pp. 733–739. https://doi.org/10.1007/s11524-010-9466-0

Rocha, N., Dias, A., Santinha, G., Rodrigues, M., Queirós, A. and Rodrigues, C. (2019). A systematic review of smart cities' applications to support active ageing. *Procedia Computer Science,* 160, pp. 306–313. https://doi.org/10.1016/j.procs.2019.11.086

Sanders C., Rogers A., Bowen R., Peter Bower P., Hirani S., Cartwright M., Fitzpatrick R., Knapp M., Barlow J., Hendy J., Chrysanthaki T., Bardsley M. and Newman P. S. (2012). Exploring barriers to participation and adoption of telehealth and telecare within the whole system demonstrator trial: A qualitative study. *BMC Health Services Research,* 12, 220.

Silvennoinen, P. and Heikkinen, S. (2023). Elderly people's perceptions of ICT's role in alleviating social isolation during the COVID-19 pandemic. *Finnish Journal of eHealth and eWelfare,* 15(1). https://doi.org/10.23996/fjhw.122270

Smets, A.J.H. (2011). Housing the elderly: segregated in senior cities or integrated in urban society? *Journal of Housing and the Built Environment,* 27(2), pp. 225–239. https://doi.org/10.1007/s10901-011-9252-7

Stefanov, D.H., Zeungnam Bien and Won-Chul Bang (2004). The smart house for older persons and persons with physical disabilities: structure, technology arrangements, and perspectives. *IEEE Transactions on Neural Systems and Rehabilitation Engineering,* [online] 12(2), pp. 228–250. https://doi.org/10.1109/TNSRE.2004.828423

Torku, A., Chan, A.P.C. and Yung, E.H.K. (2020). Implementation of age-friendly initiatives in smart cities: probing the barriers through a systematic review. *Built Environment Project and Asset Management,* 11(3), pp. 412–426. https://doi.org/10.1108/bepam-01-2020-0008

UNDESA, 2019, *World population prospects 2019.* online database.

UNECE Policy Brief on Ageing No. 24. May 2020, https://unece.org/fileadmin/DAM/pau/age/Policy_briefs/ECE_WG-1_35.pdf

van Hoof, J., Kazak, J., Perek-Białas, J. and Peek, S. (2018). The challenges of urban ageing: Making cities age-friendly in Europe. *International Journal of Environmental Research and Public Health,* 15(11), p.2473. https://doi.org/10.3390/ijerph15112473

Visvizi, A. (2023) Computers and human behavior in the smart city: Issues, topics, and new research directions, *Computers in Human Behavior,* 140, 107596, https://doi.org/10.1016/j.chb.2022.107596

WHO (2002). *Active aging: A policy framework.* Geneva, Switzerland: WHO, 2002.

Zhang, Y., Chen, G., He, Y., Jiang, X. and Xue, C. (2022). Social interaction in public spaces and well-being among elderly women: Towards age-friendly urban environments. *International Journal of Environmental Research and Public Health*, 19(2), p. 746. https://doi.org/10.3390/ijerph19020746

Zrałek, M., (2012). Kreowanie dobrych warunków mieszkaniowych i przyjaznego środowiska zamieszkania ludzi starszych. w: M. Zrałek, (ed.) *Przestrzenie starości*. Sosnowiec: Oficyna Wydawnicza Wyższa Szkoła Humanitas, pp. 87–105.

10 Smart transport systems and smart cities' growth and development

The case of Poland

Agnieszka Domańska and Radosław Malik

10.1 Introduction

The integration of smart city principles in Polish cities has been pronounced over recent years. This progression is primarily attributable to significant European Union (EU) funding (Masik et al., 2021; Godlewska-Majkowska, 2018), which aids in addressing urban challenges that could otherwise considerably impede the sustained growth of the economy and the well-being of the society (Malik & Janowska, 2019). The undeniable success in effectively deploying this financial support to modernize Polish cities and enhance the quality of life of their residents is manifest in myriad facets of the cities' daily operations (Tomal, 2020).

Although significant advancements have been realized in the technological development of urban grids and public transport infrastructure, these mask persistent challenges associated primarily with the over-density of numerous residential districts. This results in congestion and its subsequent social implications (Jonek-Kowalska & Wolniak, 2021). Such implications, including the detrimental effect on quality of life and public health, such as premature deaths (Haider et al., 2013; Levy et al., 2010) and increased travel durations, are frequently mentioned. In addition, the destructive impact of increased fuel consumption on the environment, including air and noise pollution, has been shown (Vencataya et al., 2018). A primary catalyst for these issues, particularly in many Polish cities, is that the infrastructure quality does not cater to contemporary mobility needs. Consequently, in numerous locations, solutions like intelligent transport systems (ITSs) are regarded as potential mitigations, aiming to address the deficits in city transport infrastructure and bolster the smart city evolution (Lewicki et al., 2019).

To contextualize smart city development, several endeavors have been made to operationalize its growth (Visvizi, 2022). In one approach, the initial phase, termed smart city 1.0, predominantly involves leveraging technological innovations to augment urban quality of life. The smart city 2.0 paradigm underscores the significance of quality of life, equating local governance with contemporary technology. Smart city 3.0 extends this by emphasizing community engagement and facilitating resident involvement in urban planning and

DOI: 10.1201/9781003415930-14

decision-making processes (Szarek-Iwaniuk & Senetra, 2020; Cohen, 2014; Giffinger, 2007). Consequently, the comprehensive development of smart city initiatives is increasingly recognized to depend not solely on technological interventions but also on integrating social perspectives (Visvizi et al., 2018). This emphasizes the importance of seamlessly combining technology-driven and non-technology-driven solutions to support smart city development (Lytras & Visvizi, 2018). These encompass areas such as smart society (Varela-Guzmán et al., 2021), smart service (Malik et al., 2022), smart governance (Visvizi & Troisi, 2022), and smart organizations (Godlewska-Majkowska & Komor, 2020).

However, several cities in Poland lag, particularly when juxtaposed against top-tier cities in the EU, and generally cannot be appraised within the scope of the 2.0 or 3.0 frameworks. This is primarily because residents are customarily excluded from participating in or influencing urban planning and decision-making. Decisions pertaining to residential areas are predominantly bureaucratic, and, in many instances, they are not formulated with the residents' needs and preferences in mind (Janik et al., 2023). This implies that without fulfilling the prerequisites of smart city 1.0, contemplating higher stages is untenable. Therefore, this chapter principally addresses the initial phase, as ITSs deployed in many Polish cities serve as technological interventions designed to alleviate spatial and infrastructural constraints that impede residents' quality of life.

Considering the above, the objective of this chapter is to elucidate the characteristics of urban development and the "smartification" of Polish cities, with an emphasis on transport and mobility challenges. The assessment underscores that appreciable outcomes in these domains have been largely attributable to initiatives supported by EU funds. Nonetheless, many undertakings executed in Polish cities, predominantly co-financed by the EU-backed Operational Programs, are insufficient to address the fundamental impediments to high-quality living in contemporary Polish urban environments. Notably, these impediments include excessive urban density and sprawl, both of which culminate in acute mobility constraints. Despite the considerable advancements made by numerous Polish cities, this chapter contends that certain adverse outcomes, such as traffic congestion and spatial inaccessibility (where car ownership and vehicular traffic significantly surpass the actual capacities of the intra-urban road network), have been exacerbated. Subsequent sections delve into the deployment of ITSs, advocating their potential as a solution to these growth constraints, and highlighting their capability to invigorate development and augment the spatial dimensions of economic expansion in Polish agglomerations.

The subsequent segments of this chapter are organized in the following manner. The ensuing section presents a literature review concerning ITSs in smart cities. The subsequent section delves into the primary challenges confronting Polish cities in terms of transportation and mobility. The fourth section evaluates the contribution of ITSs to the development of smart cities in Poland. This is then followed by a summary and conclusions.

10.2 Spatial development powered by intelligent transport systems in smart cities

Transitioning urban regions into thriving and sustainable cities through the integration of ITSs has been identified as a pivotal objective for local governments. This application has been extensively studied across numerous cities globally (Shamsuzzoha et al., 2021; Yigitcanlar et al., 2019). Research demonstrates that the incorporation of ICT technologies constitutes the foundation of ITS. The advancements in foundational technologies, combined with the application of cutting-edge solutions and progressive technical standards, substantially enhance the quality of ITS (Lai et al., 2020). The pivotal roles of big data (Kaffash et al., 2021), artificial intelligence (Yigitcanlar et al., 2020), the internet of things (Al-Turjman & Lemayian, 2020), and 5G networks (Gohar et al., 2021) in advancing ITS within smart cities are well-documented. Nonetheless, the deployment of such innovative solutions is intricate, often hampered by challenges including limited governmental support, financial barriers, underdeveloped transport infrastructure, and restricted technological readiness (Tran et al., 2022).

ITS within the framework of smart cities significantly affects spatial development across multiple dimensions in urban areas. However, this influence is distinct from the advancements driven by core transport infrastructure investments, such as in roads, railways, and metro systems (Grant-Muller & Usher, 2014). Evidence suggests that these systems bolster the efficient utilization of other mobility components, amplifying their advantages for urban populations (Zhao et al., 2022). Additionally, ITS offers numerous environmental, economic, and social boons, including enhanced accessibility, broader coverage, greater flexibility, improved safety, and comprehensive integration of transport processes (Butler et al., 2020). Hence, while its impact on existing spatial development attributes remains crucial, it is often circumscribed by entrenched core infrastructure.

Nevertheless, the recognition of the merits derived from ITS is increasingly informing spatial planning decisions in smart cities (Zhu, 2022). This might result in a more pronounced influence on future spatial trajectories of smart cities (Richter et al., 2020). Conversely, unchecked spatial growth without the supportive infrastructure of ICT-enabled ITS could exacerbate the adverse effects of urbanization, such as congestion, urban sprawl, heightened environmental repercussions, and other urban diseconomies (Pradhan et al., 2021).

10.3 Smart city development challenges related to mobility in Polish cities

In this section, we provide an overview of literature pertaining to the implementation of smart city concepts in Poland with insights into the financial and organizational strategies that drive the development of Polish municipalities within the smart city paradigm. The persisting challenges related to mobility will be highlighted, notwithstanding the undeniable advancements and progress achieved.

Research on smart cities in Poland discussing mobility predominantly addresses public transport and internal road infrastructure. It has been shown that while each thematic dimension of the National Municipal Policy 2023 (a strategic document endorsed by the Polish government in 2015) holds significance for a city's sustainable evolution, it is the domain of transport and mobility that most adequately mirrors the real orientations and priorities set by Polish urban policies (Gadziński & Goras, 2019). Consequently, a substantial portion of funds earmarked for agglomeration development has been directed toward transportation, with a notable emphasis on urban public transport. This includes for instance: upgrading vehicle fleets with electric buses, purchase of trams, and enhancing the metro system in Warsaw, as well as refurbishing extant infrastructural links.

Moreover, research offers a perspective on smart city progression in Poland through a mobility lens, delving into the modern road infrastructure strategies in Szczecin within the framework of the smart city development model tailored for mid-sized European cities. Upon evaluating investments in Szczecin that align with advancing smart city solutions, numerous transport and mobility projects supported by Szczecin's municipal leaders were positioned as bolstering smart city evolution (Gazińska, 2018). The other research, while refining select definitions of smart city in relation to mobility and the environment, crafted an economic model for integrating electric buses in Poland's metropolises. The model was employed to optimize battery charging cycles in buses powered electrically (Janecki & Karoń, 2014).

Nevertheless, a more nuanced analysis reveals that fostering smart city evolution, especially within the mobility sector, was not the primary intent behind the strategies of municipal authorities in Poland. Mobility has been probed in the context of coastal cities, accounting for unique challenges faced by such locales, such as heightened tourism and urban dilemmas linked to transient durations. Their scholarly endeavors aim to reconcile two avenues of research – smart cities and the blue growth concept – that had hitherto been explored in relative isolation. These authors executed proprietary studies in 10 prominent Polish cities situated alongside water bodies, assessing, inter alia, the preparedness of administrative offices to adopt smart city paradigms (Orłowski & Szczerbicki 2019).

Polish cities have undergone significant positive transformations since the early 1990s, most markedly post-2004 when Poland acceded to the European Union. Projects centered on the implementation of smart city paradigms have predominantly been funded through a diverse array of public sources, augmented substantially by the EU funds. Eminent EU-supported initiatives encompass the *Technical Assistance Operational Programme 2014–2020, EU Programme Infrastructure and Environment*, and *Intelligent Development*. Over the past two decades, Poland has emerged as the foremost beneficiary of the EU funds. Specifically, during the fiscal perspectives of 2007–2013 and 2014–2020, the EU earmarked €67.3 billion and €105.8 billion, respectively, for the Polish economy. As per the Ministry of Funds and Regional Policy, since

2004, Poland has effectively utilized approximately €154.44 billion and actualized over 273,000 projects, cumulatively valuing public subsidies at €265.2 billion. Post Poland's EU accession, investments channeled into road infrastructure have reached a staggering PLN 583 billion, with EU funds constituting PLN 349 billion thereof.

A significant quantum of these financial inflows has advantaged Polish cities, culminating in the realization of 17,427 projects. Of these, 722 initiatives were delineated as "smart city." It is pivotal to underscore that endeavors bolstering the propagation of smart city paradigms were apportioned across diverse priorities and drew financing from manifold program classifications. As an illustration, projects emphasizing ICT within urban contexts were beneficiaries of virtually all national and regional programs spanning 2014–2020. These encompassed sectors such as transport (encompassing ITSs and local traffic control centers), power engineering (highlighting intelligent energy networks), and environmental conservation (featuring automation and robotic systems, notably in sewage treatment or incineration plants). Resources for these objectives predominantly flowed from the Infrastructure and Environment Program and Regional Programs (EFRDs). In a parallel vein, projects supporting process automation and the enhancement of city-centric data exchange systems received substantial backing primarily from the Intelligent Development Program and Regional Programs. Additionally, the European Social Fund underwrote projects amplifying digital proficiencies in domains tethered to labor markets, education, and social inclusion – exemplified by the "Digital School" network program, conceived to elevate digital acumen among educators and pupils and equip schools with ICT apparatuses.

Moreover, programs that overtly champion the inculcation of smart city tenets have been instituted, such as the *"Human Smart Cities. Smart Cities co-created by residents"* program, which, with a budget allocation of €18 billion, was inaugurated in 2017. This initiative's raison d'être was to transmute urban landscapes into congenial habitats through the infusion of intelligent innovations, bolster urbanites' collective stewardship over urban territories, and galvanize their proactive engagement in its governance. The *"Human Smart Cities. Smart Cities co-created by residents"* initiative was fiscally sustained by the *Technical Assistance Operational Program 2014–2020.*

As previously highlighted, the bulk share of funds has been earmarked for investments in public transport, specifically for the modernization of both rolling stock and infrastructure. An examination of the support for the development of municipal and agglomeration transport, under programs co-financed by the European Union from 2004 to 2020, reveals that such initiatives were consistently backed in all financing perspectives by national and regional programs alike. A comprehensive summary of the projects aimed at enhancing transport infrastructure and public transport in cities is found in Table 10.1.

It is evident that both the number and the value of public transport projects co-financed by European funds have consistently risen from 2004 to 2020. During the initial programing period (2004–2006), a total of 37 such projects

Table 10.1 Support for the development of urban and agglomeration transport under programs co-financed from the funds of the European Union in the period 2004–2020

Support level	2004–2006	2007–2013	2014–2020
	Sectoral Operational Program Transport 2004–2006: Measure 1.1 "Modernization of railroad lines in relations between agglomerations urban areas and in agglomerations" (support for the development of the subway system in Warsaw)	Operational Program Infrastructure and Environment 2007–2013: (a) Measure 7.3 "Urban transport in metropolitan areas" (support for eight largest metropolitan areas: Katowice, Warsaw, Toruń and Bydgoszcz, Tricity, Kraków, Poznań, Wrocław, and Łódź) (b) Measure 8.3 "Development of Intelligent Transport Systems Transport Systems"	Operational Program Infrastructure and Environment 2014–2020: Measure 6.1 "Development of public transport public transport in cities" (support for provincial centers and their functional areas)
	Integrated Operational Program Regional development 2004–2006: Measure 1.6 "Development of transport Public transport in agglomerations" (support for seven agglomerations urban agglomerations of more than 500k inhabitants: Warsaw, Upper Silesia, Krakow, Tricity, Poznan, Wroclaw, and Lodz)	Operational Program Development of Eastern Poland 2007–2013: Measure 3.1 "Urban transport systems collective transport" (support for five provincial capitals of Eastern Poland: Lublin, Rzeszów, Białystok, Kielce, and Olsztyn)	Operational Program Poland of Eastern Poland 2014–2020: Measure 2.1 "Sustainable transport urban" (support for five capitals provinces of Eastern Poland: Lublin, Kielce, Rzeszów, Białystok, Olsztyn, and their functional areas)
Regional level	Integrated Operational Program Regional development 2004–2006: Measure 1.1 "Modernization and expansion of the regional transport system transport system" (support for cities over 50,000 inhabitants)	Regional Operational Programs 2007–2013 (support for the development of transport urban transport in the case of almost all provinces – with the exception of Warmian-Masurian)	Regional Operational Programs 2014–2020 (support for urban transport development in all provinces – including within the framework of integrated territorial investments and other territorial development instruments)

Source: Authors, based on Gadziński & Goras (2019).

were undertaken, with a combined value of PLN 3055.7 million. This escalated to 213 projects valued at 23509 million PLN in the 2007–2013 period. In the latest planning frame, specifically 2014–2020, the figures further augmented to 422 projects with a cumulative value of 26009.7 million PLN. The EU co-financing for these investments in consecutive periods was PLN 1089.1 million, 14628.8 million, and 15067.4 million, respectively (Gadziński & Goras, 2019).

Polish cities have made significant progress in evolving into more livable spaces and exhibiting increased smartness (Kędra et al., 2023), as recognized in international rankings and by numerous scholars. The role of technological advancements and augmented research and development expenditures in Poland was shown as crucial drivers in the journey toward creating smarter cities (Sikora-Fernandez, 2018). Other scholars note that the hallmark of Polish smart cities is their investment in social capital, transport, communication infrastructure, sustainable fuel sources, economic progress, and quality of life (Amistadi et al., 2023; Xia et al., 2023; Kézai et al., 2020).

However, two primary challenges persist in the transformation of Polish agglomerations into spaces that are both smarter and more conducive to their residents. These challenges are, to an extent, intertwined. The first pertains to an absence of holistic and consistent strategies by policymakers in ensuring cities develop intelligently in a spatially and organizationally harmonious manner. The second, related to a central topic of this chapter, revolves around issues related to mobility and accessibility.

First, despite substantial funding directed toward mobility projects in Poland and the frequent integration of the smart city model into development priorities, cities generally fail to adopt a holistic and multi-dimensional smart city approach. It is evident that many projects, labeled as contributing to smart city development, often manifest as isolated initiatives that do not integrate well within the larger "urban ecosystems." This disjunction impedes the adoption and implementation of intelligent and environmentally conscious solutions in Poland, a version often referred to as "smart city 1.0." This challenge is exacerbated by the absence of such solutions in local developmental strategies, the nonexistence of relevant legal frameworks, and a scarcity of established best practices (Janik et al., 2023). Research conducted in Polish cities revealed that major agglomerations exhibit a high or very high prevalence of actions aligning with the smart city concept. The issue, however, lies in their lack of systematic organization and integration (Stawasz & Sikora-Fernandez, 2015).

Second, it is undeniable that a plethora of smart city-related projects executed in Poland have yielded a myriad of positive outcomes, especially in enhancing public transport. This encompasses both infrastructural improvements, such as the introduction of new tram lines and the Warsaw metro, and the modernization of rolling stock, characterized by modern, comfortable, and eco-friendly vehicles. Additionally, strides have been made in the digitalization and smartification of cities. However, these advancements do not directly address foundational challenges that surfaced post the 1990s transition and have been intensifying notably over the past decade and a half. A core concern is that the road infrastructure, particularly within residential areas and city centers, in many Polish cities remains unresponsive to contemporary mobility and accessibility demands. This inadequacy hampers the potential of these agglomerations to be regarded as amiable, livable spaces, let alone labeled as "smart." The crux of this predicament is rooted in the fact that a significant proportion of urban roads were designed and constructed during the 1970s and 1980s, predicated on a low automobile ownership rate and modest traffic. Consequently, this infrastructure proved ill-equipped to accommodate the evolving mobility preferences and surge in car ownership that has been characteristic of the rapidly expanding urban populations from the early 1990s to the present.

Overpopulation in numerous residential districts, engendering pronounced mobility challenges, stems directly from significant deficiencies in spatial planning. This has permitted indiscriminate and unchecked urban development and the acquisition of inter-city areas by private construction entities. This detrimental trend also correlates with the previously mentioned issue – decision-makers' inability to establish coherent urban development policies, particularly concerning spatial planning. The most acute consequences of this urban "densification" and unchecked suburban sprawl include infrastructural bottlenecks accompanied by escalating congestion and mobility restrictions (Gadziński & Goras, 2019). An excessive reliance on automobile travel yields numerous repercussions, diminishing residents' quality of life and causing tangible economic losses.

Among other factors, a predominant cause of this situation appears to be the absence of a sufficient legal framework, especially a dearth of local spatial development plans. Such plans are pivotal tools that support construction activity in a structured fashion (Howe & Langdon, 2002). This phenomenon, notably evident in many Polish cities and especially in Warsaw, poses significant spatial challenges, as documented in both scholarly literature and international reports. Scholars highlight difficulties in restraining haphazard construction, which leads to environmental disarray and impedes cities' sustainable growth (Hajduk, 2015). The inconsistency in urban development decisions in Poland has resulted in numerous cities expanding in a disorganized manner (Hajduk 2015, p. 183). Gadziński and Goras contend that urban sprawl, coupled with unregulated suburban growth, has magnified travel between central and neighboring rural areas. This has exacerbated road congestion, elevating noise and pollution levels (Gadziński & Goras, 2019, p. 36).

Increased traffic congestion, stemming from excessive urbanization and burgeoning urban populations (key spatial facets of urban operation), in Polish cities is noted in numerous indices and reports. For instance, the TomTom Traffic Index 2020 ranks Łódź and Kraków among the top 10 most congested European cities, surveying 416 cities across 57 countries. Łódź residents, for instance, endure 47% additional travel time in traffic compared to free-flow scenarios. Traffic conditions in other Polish cities, including Kraków (45%), Poznań

(44%), Warsaw (40%), and others, similarly deteriorate. This index reveals a traffic volume uptick in 2019 across all Polish cities from the previous year.

Furthermore, the economic ramifications of traffic congestion have been accentuated in various studies. Wrocław and Łódź are Poland's most grid-locked city, with an average driver spending approximately 235 and 211 hours annually in traffic, respectively, followed closely by Kraków, Szczecin, and Warszawa (Tomtom, 2022). Nevertheless, other international indices, such as the Cities in Motion by the IESE Business School's Center for Globalization and Strategy, offer a more moderate perspective when gauging these challenges on a global scale. In this analysis, which considers over 150 cities globally, two Polish cities, i.e., Warsaw and Wrocław, have demonstrated steady improvement in recent years relative to their global counterparts (IESE, 2022).

In conclusion, the positive urban transformations in Poland in alignment with smart city principles over recent years are juxtaposed with pronounced shortcomings, highlighting the absence of a holistic approach encompassing essential components like legislation, infrastructure, and technology. These trends underscore the need for a comprehensive perspective on urban "smartness" that emphasizes not just ICT-driven processes inherent to smart city evolution but also the crucial economic and spatial considerations vital for a city's progression to a "smart" status and overall development. This underlines an urgent necessity to devise strategies addressing these issues sustainably, with mobility challenges from excessive traffic standing as a paramount concern.

10.4   Intelligent transport systems contribution to smart city development in Poland

Given the magnitude of transport infrastructure challenges, the transformation of Polish metropolises and medium-sized towns in line with the principles of the smart city paradigm necessitates solutions that empower cities to progress, notwithstanding the physical constraints arising from traffic volumes that exceed road infrastructure capacities. Such advancements could be realized by endowing local communities with substantial mobility potential, rooted in novel, eco-friendly, and energy-efficient technologies.

ITSs often incorporate cutting-edge technological solutions and epitomize the core tenets of the smart city ideology. Specifically, by enhancing the fluidity of vehicular movement within road networks – achieved, for instance, via automatic detection mechanisms that prioritize public transport and cyclists – ITS promotes the adoption of greener transportation modalities and aids in curbing vehicular emissions. Consequently, cities that integrate ITS can optimize their road infrastructure utility, refine vehicular traffic management, and expedite public transportation services (Kaffash et al., 2021).

ITSs are multifaceted, typically comprising various modules (refer to Table 10.1). Crucial components of the system cater to the provision of real-time data for both drivers and passengers, aiding them in both journey planning and execution. Conversely, another subset focuses on ensuring the safety of road users. Distinct modules, for example, are tailored to environmental monitoring and conservation, frequently integrating machinery that monitors atmospheric quality (Costabile & Allegrini 2008).

It has been underscored that ITS encompasses an array of technologically sophisticated solutions. For illustrative purposes, we examine an ITS variant furnished by Siemens – a firm that has spearheaded a significant portion of ITS projects in Poland (consult Table 10.2). The system under scrutiny incorporates tools designed for orchestrating and overseeing traffic signaling (SitrafficScala), the 1 Watt technology (Sitraffic One), and motion. The former is an integrated, modular signaling design and execution software encompassing visualization modules for signaling group states, detector readings, and public transport vehicle reports, facilitating the analysis and refinement of signaling protocols. This system affords insights into several metrics, such as fluctuations in travel durations, their distribution, and signal waiting times. Another utility, termed "Motion" (Method for Optimization of Traffic signals in Online controlled Networks), embodies an adaptive traffic network control mechanism that assimilates data from diverse sources. Leveraging information harvested from strategically positioned detectors (e.g., at pivotal intersections), this management module delineates the prevailing traffic landscape and projects subsequent shifts. The manufacturer accentuates that this system amplifies the efficacy of pre-existing infrastructure, given that traffic enhancements are realized network-wide, as opposed to merely at isolated junctions – this is especially palpable during rush hours, marked by a decline in stops attributable to traffic signals.

Another form of ITS commonly termed as organized parking management encompasses a suite of tools designed to alleviate the challenge of searching for parking, which inadvertently contributes to extraneous traffic (Berenger Vianna et al., 2004). This adaptable guidance system for available parking spaces offers real-time data, detailing factors such as the occupancy rate of parking spaces, anticipated wait times for an open slot, the total count of vacant spaces, and pertinent public transport links. In addition, an integrated reservation module can enable users to secure parking slots online (Gangwani & Gangwani, 2021).

Table 10.2 Main goals set for ITS projects in Poland

1. Regulating vehicular traffic within the road network to alleviate congestion.
2. Optimizing the utilization of extant road and transportation infrastructure.
3. Enhancing travel conditions with a consideration for transport intermodality.
4. Accelerating public transport, with an emphasis on trams.
5. Promoting greater reliance on public and alternative modes of transportation.
6. Disseminating timely information beneficial for both drivers and passengers during travel planning and execution.
7. Augmenting safety measures for all road users.
8. Overseeing and safeguarding the environment.

Source: Authors, based on Siemens' projects.

ITS implementations are not exclusively tailored for the aid of drivers and public transport users; they can also bolster the mobility of cyclists. Such innovations grant priority to cyclists and, by augmenting the efficacy of oversight across cycling corridors, they heighten the allure of bicycles as a viable transportation mode, concurrently facilitating emission reduction. A comprehensive overview of the ITS initiatives instituted in Polish cities can be found in Table 10.3.

Table 10.3 Chosen ITS solutions implemented in Polish cities

ITS solution	Functionality	Effects/benefits	Modules/additional information
Sitraffic One – 1 Watt Technology	Solutions for traffic signals: controllers, traffic signal, and pedestrian push buttons acoustic devices	Less energy consumption, less air pollution, and saving money; 80% reduction in energy consumption compared to a 40V LED solution; savings of up to €500,000 per year for a city the size of Berlin; a pilot installation in the city of Herne provided the city with a 19,000 kWh reduction in energy consumption compared to incandescent bulbs	1 Watt technology Silux2 VLP Beacons
Siemens CyAM/ CyPT air monitoring	Air monitoring in terms of key indicators; generation of hourly, daily, and weekly reports; trend analysis and historical comparisons; basis for decision-makers to take concrete actions now and in the future; dynamic traffic management; based on historical emissions data combined with data from other sources (weather, events, traffic volume, energy consumption, etc.); Flexible parking fees; database for considering the creation of low-emission zones	Understanding how an agglomeration/ city complies with EU regulations; understanding the socio-economic impacts associated with poor air quality; proven accuracy level of 90% for 5-day prediction; improved air quality through systems integration and dynamic traffic management; improving the quality of life of residents and the image of the city, avoiding fines.	Artificial intelligence based on neural network techniques used for prediction; continuous collection of street network air quality data depending on traffic volume analysis; variety of measures: from improving traffic flow and intelligently shifting emission loads to less critical areas, to establishing dynamic environmental zones; dynamic restrictions on the entry of heavy and diesel vehicles into the zones; dynamic speed limits; automatic or manual response to the situation or forecasts

(Continued)

Table 10.3 (Continued)

ITS solution	Functionality	Effects/benefits	Modules/additional information
SITRAFFIC PLATFORM	Traffic management and control in the city; priority for public transport; management of parking lots; maintenance and monitoring; data integration; intuitive, transparent visualization of signaling programs, intersections, and "green wave."	Various effects and benefits depending on the module: less pollution, less congestion, less time wasted by the traffic users by support for individual traffic, etc.	Visualization module; quality management module; strategic management module; support for public transport; intermodal transportation; parking module
Sitraffic MOTIONS (Method for Optimization of Traffic signals in Online controlled Networks/On-line Network Optimization Method using traffic signals)	On the basis of data collected from detectors located in appropriate places, SITRAFFIC MOTION generates the current traffic situation and predicts the situation that may arise	Measurable improvements in traffic flow without making significant changes to the existing infrastructure	

Source: Own research.

ITS components have been incorporated in cities such as Warsaw, Kraków, Poznań, Rzeszów, and Białystok to address infrastructure constraints. Up to 2017, Polish cities allocated approximately 580 million PLN to ITS-related initiatives, which benefited considerably from EU funding. These systems facilitate tangible enhancements in traffic flow without necessitating extensive modifications to the prevailing infrastructure. Consequently, they positively impact not only residents' mobility but also air quality, thus elevating the overall quality of life for inhabitants.

In conclusion, the deployment of ITS presents a potent solution to the spatial challenges and resultant mobility impediments confronting the growth of smart cities in Poland. Systems grounded in contemporary ICT are poised to play an indispensable role in mitigating prevalent urban issues, such as communication gridlocks, traffic congestion, and parking scarcity. As such, ICT-driven ITSs appear to possess the potential to shape both linear and non-linear trajectories of urban growth and evolution.

10.5   Summary and conclusions

The literature review concerning the smart city concept within the Polish context reveals that research primarily focuses on three central areas: measurement, governance, and mobility. Investigations assessing the extent of smart city concept implementation in Poland typically indicate a moderate level. This modest implementation is frequently attributed to an absence of an integrated urban approach, as highlighted by studies examining smart cities from a governance standpoint. Notably, our analysis underscores that mobility emerges as a pivotal research area and a significant implementation domain within Polish smart cities, supported extensively by programs co-financed by EU funds. Additionally, a growing consensus among scholars suggests that the application of smart city concepts in Poland should more aptly be viewed as foundational or pilot initiatives, rather than comprehensive policies.

Our review further accentuates the necessity of adopting a holistic approach to create intelligent and conducive urban spaces for enhanced living. A mature smart city policy should pragmatically amalgamate all essential elements: tangible infrastructure paired with appropriate public transport and logical residential development strategies (i); proper legislative frameworks guiding urban development, encompassing both the planning and management of urbanization activities (ii); and the incorporation of smart city concepts targeting intangible factors that elevate urban life quality (iii).

Given the plethora of initiatives aimed at enhancing Poland's urban spaces juxtaposed with palpable governance shortcomings, it's evident that a comprehensive methodology incorporating all three crucial elements is lacking. While many Polish urban areas have met an increasing number of conditions stipulated by smart city principles – largely owing to dedicated EU funding – legislative and spatial planning deficiencies, coupled with tangible infrastructure gaps adversely affecting mobility, have resulted in considerable challenges, notably traffic congestions.

Nevertheless, as delineated in our discourse, ITSs, by enhancing flow and accessibility in mobility and endorsing more sustainable transport methods, can play an instrumental role in narrowing the disparity between smart city tenets and the urban quality of life in Poland. Consequently, the integration of ICT within these systems profoundly reshapes urban spatial dynamics, invigorating infrastructure-related revitalization processes.

The discussed cases of ITS implementation in Poland examine efforts to augment the sustainability of smart cities and territories by addressing regional development non-linearities, thus refining the spatial composition of smart cities. The findings underscore the prevalence of ICT-centric solutions in smartification discussions, overshadowing a more holistic amalgamation of non-technological and ICT-oriented strategies among policy and business leaders. The text elucidates the societal implications of this narrow focus concerning the spatial design of smart cities. The chapter contends that a harmonized blend of both non-technological and technology-driven solutions yields the most beneficial results for urban spatial development.

Acknowledgments

Research presented in this chapter are related to research conducted in the framework of the National Science Centre (NCN) grant "Smart Cities: Modelling, Indexing and Querying Smart City Competitiveness" (Nr DEC-2020/39/B/HS4/00579), led by Anna Visvizi.

References

Al-Turjman, F & Lemayian, JP 2020, 'Intelligence, security, and vehicular sensor networks in internet of things (IoT)-enabled smart-cities: An overview', *Computers & Electrical Engineering*, vol. 87, p. 106776. https://doi.org/10.1016/J.COMPELECENG.2020.106776

Amistadi, L, Bradecki, T, & Uherek-Bradecka, B 2023, 'Resilient university campus in the city in COVID and post-COVID era—Recommendations, guidelines, and evidence from research in Italy and Poland', *Urban Design International*, vol. 28, no. 2, pp. 141–151. https://doi.org/10.1057/s41289-022-00211-y

Berenger Vianna, MM, da Silva Portugal, L & Balassiano, R 2004, 'Intelligent transportation systems and parking management: Implementation potential in a Brazilian city', *Cities*, vol. 21, no. 2, pp. 137–148. https://doi.org/10.1016/J.CITIES.2004.01.001

Butler, L, Yigitcanlar, T & Paz, A 2020, 'How can smart mobility innovations alleviate transportation disadvantage? Assembling a conceptual framework through a systematic review', *Applied Sciences*, vol. 10, no. 18, p. 6306. https://doi.org/10.3390/APP10186306

Cohen, B 2014, 'The 10 smartest cities in Europe', *Fast Company*. Available at: https://www.fastcoexist.com/3024721/the-10-smartest-cities-in-europe [Accessed: 4 April 2023].

Costabile, F & Allegrini, I 2008, 'A new approach to link transport emissions and air quality: An intelligent transport system based on the control of traffic air pollution', *Environmental Modelling & Software*, vol. 23, no. 3, pp. 258–267. https://doi.org/10.1016/J.ENVSOFT.2007.03.001

Gadziński, J & Goras, E 2019, *Raport o stanie polskich miast, Transport i mobilność miejska*, Instytut Rozwoju Miast i Regionów, Warszawa.

Gangwani, D & Gangwani, P 2021, 'Applications of machine learning and artificial intelligence in intelligent transportation system: A review', *Lecture Notes in Electrical Engineering*, vol. 778, pp. 203–216. https://doi.org/10.1007/978-981-16-3067-5_16/COVER

Gazińska, O 2018, 'Changes in road infrastructure in smart city development model of Szczecin, Poland', *Przestrzeń i Forma*, vol. 36, pp. 177–190.

Giffinger, R 2007, *European Smart City Model (2007–2015)*. Vienna University of Technology. Available at: http://www.smart-cities.eu [Accessed: 2 August 2023].

Godlewska-Majkowska, H 2018, 'Polarity of the regional space–The dilemma of shaping socio-economic development in Poland', *Miscellanea Geographica*, vol. 22, no. 2, pp. 99–109. https://doi.org/10.2478/mgrsd-2018-0011

Godlewska-Majkowska, H & Komor, A 2020, 'The role of intelligent organisations in creating favourable conditions for the development of entrepreneurship', *European Research Studies*, vol. 23, pp. 897–922.

Gohar, A, Nencioni, G, Khyam, O & Li, X 2021, 'The role of 5G technologies in a smart city: The case for intelligent transportation system', *Sustainability*, vol. 13, no. 9, pp. 5188. https://doi.org/10.3390/SU13095188

Grant-Muller, S & Usher, M 2014, 'Intelligent transport systems: The propensity for environmental and economic benefits', *Technological Forecasting and Social Change*, vol. 82, no. 1, pp. 149–166. https://doi.org/10.1016/J.TECHFORE.2013.06.010

Haider, M, Kerr, K & Badami, M 2013, 'Does Commuting Cause Stress? The Public Health Implications of Traffic Congestion'. Available at: https://ssrn.com/abstract=2305010 or http://doi.org/10.2139/ssrn.2305010

Hajduk, S 2015, 'The spatial management vs. Innovativeness of Medium-Size Cities of Poland', *Procedia - Social and Behavioral Sciences*, vol. 213, pp. 879–883. https://doi.org/10.1016/j.sbspro.2015.11.499

IESE 2022. Cities in motion index 2022. Available at: https://www.iese.edu/stories/smart-sustainable-cities-in-motion-index/ [Accessed: 7 August 2023]

Janecki, R & Karoń, G 2014, 'Concept of smart cities and economic model of electric buses implementation', in J Mikulski (ed.), *Telematics - Support for Transport. TST 2014*. Communications in Computer and Information Science, vol 471. Berlin, Heidelberg: Springer. https://doi.org/10.1007/978-3-662-45317-9_11

Janik, A, Ryszko, A & Szafraniec, M 2023, 'Intelligent and environmentally friendly solutions in smart cities' development—Empirical evidence from poland', *Smart Cities*, vol. 6, no. 2, pp. 1202–1226. https://doi.org/10.3390/smartcities6020058

Jonek-Kowalska, I & Wolniak, R 2021, 'Economic opportunities for creating smart cities in Poland. Does wealth matter?', *Cities*, vol. 114, p. 103222.

Kaffash, S, Nguyen, AT & Zhu, J 2021, 'Big data algorithms and applications in intelligent transportation system: A review and bibliometric analysis', *International Journal of Production Economics*, vol. 231, p. 107868. https://doi.org/10.1016/J.IJPE.2020.107868

Kędra, A, Maleszyk, P & Visvizi, A 2023, 'Engaging citizens in land use policy in the smart city context', *Land Use Policy*, vol. 129, p. 106649. https://doi.org/10.1016/j.landusepol.2023.106649

Kézai, PK, Fischer, S, & Lados, M 2020, 'Smart economy and startup enterprises in the Visegrád Countries – A comparative analysis based on the Crunchbase Database', *Smart Cities*, vol. 3, no. 4, pp.1477–1494. https://doi.org/10.3390/smartcities3040070

Lai, CS, Jia, Y, Dong, Z, Wang, D, Tao, Y, Lai, QH, Wong, RTK, Zobaa, AF, Wu, R & Lai, LL 2020, 'A review of technical standards for smart cities', *Clean Technologies*, vol. 2, no. 3, pp. 290–310. https://doi.org/10.3390/CLEANTECHNOL2030019

Levy, JI, Buonocore, JJ & von Stackelberg, K 2010, 'Evaluation of the public health impacts of traffic congestion: A health risk assessment', *Environmental Health*, vol. 9, art. no. 65.

Lewicki, W, Stankiewicz, B & Olejarz-Wahba, AA 2019, 'The role of intelligent transport systems in the development of the idea of smart city', in *Scientific and Technical Conference Transport Systems Theory and Practice* (pp. 26–36). Cham: Springer International Publishing.

Lytras, MD, & Visvizi, A 2018, 'Who uses smart city services and what to make of it: Toward interdisciplinary smart cities research'. *Sustainability*, vol. 10, no. 6, p. 1998.

Malik, R & Janowska, AA 2019, 'The next 100 years – Applying megatrends to analyze the future of the Polish economy', *Nierówności Społeczne a Wzrost Gospodarczy*, vol. 1, no. 57, pp. 119–131. [Online] https://doi.org/10.15584/nsawg.2019.1.8

Malik, R, Visvizi, A, Troisi, O & Grimaldi, M 2022, 'Smart services in smart cities: Insights from science mapping analysis', *Sustainability*, vol. 14, no. 11, pp. 6506.

Masik, G, Sagan, I & Scott, JW 2021, 'Smart city strategies and new urban development policies in the Polish context', *Cities*, vol. 108, p. 102970.

Orłowski, A & Szczerbicki, E 2019, 'Smart blue cities', *Europa XXI*, vol. 36, pp. 77–88. https://doi.org/10.7163/Eu21.2019.36.7

Pradhan, RP, Arvin, MB & Nair, M 2021, 'Urbanization, transportation infrastructure, ICT, and economic growth: A temporal causal analysis', *Cities*, vol. 115, p. 103213. https://doi.org/10.1016/J.CITIES.2021.103213

Richter, A, Löwner, MO, Ebendt, R & Scholz, M 2020, 'Towards an integrated urban development considering novel intelligent transportation systems: Urban Development Considering Novel Transport', *Technological Forecasting and Social Change*, vol. 155, p. 119970. https://doi.org/10.1016/J.TECHFORE.2020.119970

Shamsuzzoha, A, Niemi, J, Piya, S & Rutledge, K 2021, 'Smart city for sustainable environment: A comparison of participatory strategies from Helsinki, Singapore and London', *Cities*, vol. 114, p. 103194. https://doi.org/10.1016/J.CITIES.2021.103194

Sikora-Fernandez, D 2018, 'Smarter cities in post-socialist country: Example of Poland', *Cities*, vol. 78. https://doi.org/10.1016/j.cities.2018.03.011

Szarek-Iwaniuk, P & Senetra, A 2020, 'Access to ICT in Poland and the co-creation of urban space in the process of modern social participation in a smart city – A case study', *Sustainability*, vol. 12, no. 5, p. 2136.

Tomal, M 2020, 'Moving towards a smarter housing market: The example of Poland', *Sustainability*, vol. 12, no. 2, p. 683.

Tomtom 2022, TomTom traffic index, ranking 2022. Available at: https://www.tomtom. com/traffic-index/ranking/ [Accessed: 2 August 2023]

Tran, CNN, Tat, TTH, Tam, VWY & Tran, DH 2022, 'Factors affecting intelligent transport systems towards a smart city: A critical review'. https://doi.org/10.1080/15 623599.2022.2029680

Varela-Guzmán, E, Mora, H & Visvizi, A 2021, 'Exploring the differentiating characteristics between the smart city and the smart society models', *The International Research & Innovation Forum* (pp. 313–319). Cham: Springer.

Vencataya, L et al. 2018, 'Assessing the causes & impacts of traffic congestion on the society, economy and individual: A case of Mauritius as an emerging economy', *Studies in Business and Economics*, vol. 13, no. 3, pp. 230–242. https://doi.org/10.2478/ sbe-2018-0045

Visvizi, A 2022, vol. 140, *Computers in Human Behavior*, p. 107596.

Visvizi, A, Lytras, MD, Damiani, E & Mathkour, H 2018, 'Policy making for smart cities: Innovation and social inclusive economic growth for sustainability', *Journal of Science and Technology Policy Management*, vol. 9, no. 2, pp. 126–133.

Visvizi, A & Troisi, O 2022, 'Effective management of the smart city: An outline of a conversation', in *Managing Smart Cities: Sustainability and Resilience through Effective Management* (pp. 1–10). Cham: Springer International Publishing.

Xia, L, Semirumi, DT, & Rezaei, R 2023, 'A thorough examination of smart city applications: Exploring challenges and solutions throughout the life cycle with emphasis on safeguarding citizen privacy', *Sustainable Cities and Society*, vol. 98, p. 104771. https://doi.org/10.1016/j.scs.2023.104771

Yigitcanlar, T, Desouza, KC, Butler, L & Roozkhosh, F 2020, 'Contributions and risks of artificial intelligence (AI) in building smarter cities: Insights from a systematic review of the literature', *Energies*, vol. 13, no. 6, p. 1473. https://doi.org/10.3390/ EN13061473

Yigitcanlar, T, Han, H, Kamruzzaman, M, Ioppolo, G & Sabatini-Marques, J 2019, 'The making of smart cities: Are Songdo, Masdar, Amsterdam, San Francisco and Brisbane the best we could build?', *Land Use Policy*, vol. 88. https://doi.org/10.1016/J. LANDUSEPOL.2019.104187

Zhao, C, Wang, K, Dong, X & Dong, K 2022, 'Is smart transportation associated with reduced carbon emissions? The case of China', *Energy Economics*, vol. 105. https:// doi.org/10.1016/J.ENECO.2021.105715

Zhu, W 2022, 'A spatial decision-making model of smart transportation and urban planning based on coupling principle and Internet of Things', *Computers and Electrical Engineering*, vol. 102. https://doi.org/10.1016/J.COMPELECENG.2022.108222

Stawasz, D, & Sikora-Fernandez, D 2015, Management in Polish cities in accordance with the smart city concept. *Warszawa, Placet.*

Howe, J, & Langdon, C 2002, 'Towards a reflexive planning theory', *Planning Theory*, vol. 1, no. 3, pp. 209–225.

11 Autonomous vehicles in smart cities

Challenges pertaining to autonomous and connected transport. The case of Romania

Liliana Andrei, Oana Luca, and Emanuel Răuță

11.1 Introduction

Technological progress in autonomous and connected transport (ACT) is expected to have a major impact on smart city development, changing mobility behavior, the transport system, and population distribution in urban environments. In recent years, a growing body of research has focused on how to balance the potential benefits of autonomousvehicles (AVs) with the expected shortcomings (Bösch et al., 2021; Clewlow & Mishra, 2017; Kane & Whitehead, 2017). In response, there is general consensus that autonomous vehicles (AVs) will change approaches in the spatial development of cities (Agriesti et al., 2020; Fraedrich et al., 2019; Gyergyay et al., 2019). Most of the research is oriented toward social acceptance (Andorka & Rambow-Hoeschele, 2020; Bissell et al., 2020; Cavoli et al., 2017), physical and digital infrastructure (Almeaibed et al., 2021; Arfiansyah & Han, 2021; González-González et al., 2020; Hunter, 2021; Johnson & Nica, 2021), shared mobility services (An et al., 2019; Barbour et al., 2019; Mo et al., 2021), and smart city (Bezai et al., 2020; Campisi et al., 2021; Cassandras, 2017), but less is focused on spatial planning. This could be a consequence of uncertainty and lack of policies and regulation that envisage the deployment of AVs in urban environments. In addition, the studies cited above are based on different scenarios on how to tackle the disadvantages such as traffic congestion, increased energy consumption, curb management, increased travel mileage, and urban sprawl, with most of the research considering the stage when all vehicles will be autonomous, without considering the transition to full autonomy. This may affect the decision of authorities to develop spatial development or sustainable urban mobility strategies to support AVs adoption.

In order to contribute to the existing body of knowledge on AVs, to provide policymakers and urban and mobility planners with information on the potential impact of AVs on urban structure, and to further advance the research on AVs in Romania (Andrei, Luca, et al., 2022; Andrei, Negulescu, et al., 2022; Barabás et al., 2017; Carabulea et al., 2022; Ionescu et al., 2021; Luu et al., 2019; Pauca et al., 2021; Perișoară et al., 2022; Trasnea et al., 2019), the following research questions are proposed: (a) Are ACT and AVs considered in the

DOI: 10.1201/9781003415930-15

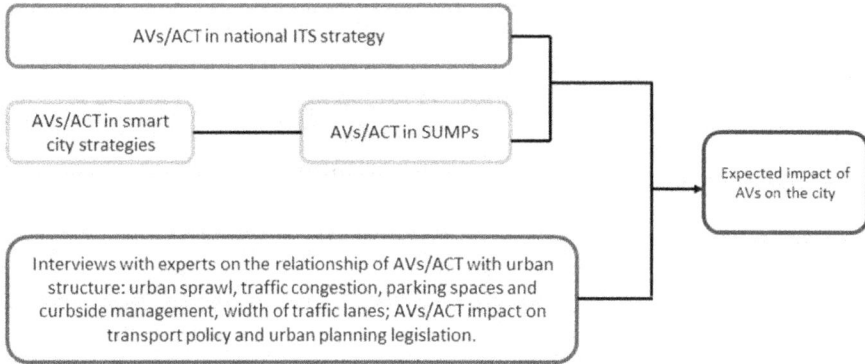

Figure 11.1 Research model for exploring the connection between AVs/ACT and urban development.

Source: Authors.

national ITS strategy and/or in the local development strategies of Romanian cities (smart cities strategy and sustainable urban mobility plans (SUMPs))? (b) What do Romanian experts actually think about the impact that ACT/AVs could have on the urban structure? (c) Is this anticipated impact similar to that expected at international level? To answer these questions, three research objectives are proposed: (1) to examine the relevant national and local urban development plans and strategies to determine whether they contain references to ACT and AVs, responding to research question (a); (2) to investigate, via interviews, experts opinion on the impact of AVs/ACT on urban structure addressing the research question (b); and (3) to discuss the results, tackling research question (c).

Considering the absence of research data on the interaction between the AV transport system and the urban space in general, starting from the micro-level (the neighborhood) to the macro-level (the city and functional urban area), and moving beyond the street–city–sidewalk–parking relationship, our chapter stands out as the first exploratory study on the connection between AVs/ACT and urban development in Romania (Figure 11.1). The chapter is structured as follows: the literature review is followed by the methodology sections. Discussion and conclusions ensue.

11.2 Literature review

AVs have a high potential to change mobility behavior as they can positively influence travel time and costs (Correia et al., 2019; Medina-Tapia & Robusté, 2019), as time spent in an AV will be used more efficiently. Different simulations have shown that by using shared AVs, up to eight conventional vehicles can be replaced (Hamadneh & Esztergár-Kiss, 2019). Using a dynamic allocation for vehicle sharing and increasing the car occupancy to four people

(Alonso-Mora et al., 2017) will reduce the number of vehicle-miles travelled and the waiting time. In addition, shared AVs transport is expected to have an important social impact. Its adoption will increase the mobility of people with disabilities (Bennett et al., 2019; Hwang et al., 2020; Kassens-Noor et al., 2021), those who cannot drive (Harper et al., 2016; Nastjuk et al., 2020), people with low incomes (Eppenberger & Richter, 2021), and those living in suburban/regional areas (Hinderer et al., 2018). Thus, the volume of car traffic is expected to increase as people tend to be more mobile (Naumov et al., 2020). Similar results are coming from different authors (Galich & Stark, 2021; Zhang et al., 2018), and, moreover, it is estimated that if AVs were to become affordable for the middle- and high-income population, they could significantly increase the number of vehicle-miles travelled. Considering this aspect, an attempt was made to assess the impact on the urban street network.

Studies have shown, however, that increased use efficiency should be expected through higher capacity, with simulations showing the possibility of increasing the capacity of a traffic lane by up to 40% (Park et al., 2021). Due to increased lateral control and increased safety of AVs, it is estimated that lane widths can be reduced by 20% (Dennis et al., 2017). Consequently, maximizing lane capacity will allow the reduction of the number of lanes. As a result, the space can be converted into bicycle lanes or it could be allocated to pavements, green spaces, and playgrounds (Stead & Vaddadi, 2019). Moreover, in low-speed zones, lane widths could be less than 2.5 m (Kisner et al., 2019), thus gaining more public space. In terms of urban space, AVs will presumably influence the location and size of parking spaces, even if these vehicles are privately owned or shared (Ostermeijer et al., 2019). The advanced features of AVs will generate a reduction in the parking space used by individual cars by an average of 62% (Nourinejad et al., 2018), as vehicles will be able to self-park (Gavanas, 2019), without requiring additional space to open doors, thus the space between vehicles being considerably reduced. In addition, the required parking spaces will be able to be located further away from home/business/service destination (Millard-Ball, 2018, 2019), given their ability to move without a driver. Parking spaces will need to be equipped with charging stations and possibly maintenance facilities (Duvall et al., 2019). The improved organization of parking spaces and the prohibition of on-street parking will increase the productivity of curbside space utilization (De Lara, 2020). This will only have a drop-in/drop-off function (Crute et al., 2018) and should be regulated by the type of service: public transport, ridesharing, delivery, etc. Given the prognosed impact of ACT, one should expect a radical change in the utilization of urban space. In more pessimistic scenarios, due to increased social inclusion and more efficient use of time spent in the vehicle, there is a significant risk of increased suburbanization or urban sprawl (Papa & Ferreira, 2018; Zhou et al., 2021).

It seems that land-use planning policies are being overtaken by rapid technological change and even by transport-related legislation. Several countries have adopted legislation regulating the operation of vehicles with different levels of autonomy (BMDV, 2022; IIHS-HLDI, 2023; Kester, 2022; Lowenberg-DeBoer et al., 2022; UK Statutory Instruments, 2022), and, in August 2022, the European Commission adopted legislation for vehicles with autonomous driving systems (ADSs) (Union, 2022b)). In addition, some countries have developed national strategies related to the deployment of AVs (FMTDI, 2015; NHTSA, 2016; NITPA, 2017), but these strategies are more oriented toward the digital side of deployment and test conditions. Although policy documents encourage actions to deploy AVs in urban areas (Union, 2020), and research projects have developed guidelines in this regard (Backhaus et al., 2019), very few authorities have taken this into account in their planning documents, be it GUPs, sustainable urban mobility plans, smart cities strategies, etc. (Andrei & Luca, 2022).

In Romania, according to regulations regarding the identity card of vehicles established by the Ministry of Transport, AVs would fall into the category of "special vehicle (special purpose vehicle) …. having specific technical characteristics which aim to execute a function that asks for special adaptations and/or equipment." However, the regulatory framework does not allow for AVs operation.

11.3 Methodology

Twenty-five Romanian cities have been selected (Figure 11.2) for the purpose of our analysis. The following set of criteria was included: the region (minimum one town in each existing development region), the number of inhabitants (2019 as the year of reference – with more than 100,000 inhabitants) according to the National Institute of Statistics, the function of the county residence municipality, and the existence of smart mobility actions/projects.

Figure 11.2 The location of the 25 cities on the Romanian map.

Source: Elaborated by authors by the use of QGIS, based on *OpenStreetMap* (2021).

Table 11.1 Background and experience of the experts interviewed (n = 20)

Characteristic		%
Professional profile	Urban/spatial planning	50
	Transportation planning	15
	Local administration	35
	ITS expert	5
	Other	5
Degree of knowledge about ACT/AVs	I have previously heard/read about this topic	60
	I participated in a panel discussion on this topic	25
	I have researched/have held a specific conference on the topic/I have published some work on the topic	10
	None the above	5

It has been examined if smart mobility-related projects are foreseen in their SUMPs or smart city strategies. The city's smart mobility initiatives were selected if official plans and documents included projects aiming to finance charging infrastructures for electric vehicles, smart traffic management, smart platforms for public transport, smart stations, smart parking, and smart streets.

A number of 20 interviews were carried out with experts in the fields of ITS, urban and transport planning, and local government focusing with priority on topics such as urban policy strategic and planning framework, relationship between AVs and urban structure, analysis of barriers to the implementation of ACT in urban areas, and potential measures to overcome them. The interviews also included five questions on Avs' future development in Romanian. Interviews were conducted between September and November 2022 among experts with extensive experience in urban mobility (Table 11.1), well acquainted with the possible impacts of introduction of AVs, and knowledge coming rather from complementary fields of research than from the AV-specific research field.

11.4   Findings and discussions

Romanian National ITS Strategy 2022–2030 which has recently been adopted (MTI, 2022) sets up objectives to introduce ACTs: (a) the development of cooperative, connected, and autonomous mobility services; and (b) the development of the technological and institutional framework for the provision of cooperative, connected, and autonomous mobility-specific ITS services. In addition, the Action Plan foresees measures related to (i) an AV pilot project (road/rail) for the development of supporting infrastructure components for AVs; (ii) modification of the National Road Traffic Code to allow testing for AV; (iii) connected vehicle training programs. However, the approach for AVs is equally well oriented toward digital infrastructure and data communication and the document does not indicate how the ACT elements are to be integrated into SUMPs. In supporting the strategic objectives, Romanian authorities have also submitted for adoption a legal amendment in the form of the Government Emergency Ordinance no. 195/2002 regarding traffic on public roads, which aims at "creating the legal premises for the authorization for testing vehicles with an autonomous driving system, for the purpose of testing the vehicles in question, on certain routes established in advance by the traffic police."

Among the 25 cities analyzed with more than 100,000 inhabitants, country residence municipalities and capital city of the country, only 14 of them have foreseen smart mobility projects in their adopted strategies (SUMPs (Andrei & Luca, 2022) or smart city strategies), mostly related to electrical vehicles charging infrastructure, smart stations, and smart traffic management. Interestingly, smart mobility projects were proposed either in SUMP or in smart cities strategies and never in both, which provides evidence for a certain lack of correlation between strategic documents, an element which was also emphasized in previous articles (Luca et al., 2021). Results indicate that the cities from the South-East Region (Galati, Braila, Constanta) have not included information on smart mobility projects, the capital city, Bucharest, does not have an official smart city strategy, and its SUMP drafted in 2017 has made no mention about AVs or ACT and only three cities, Cluj, Timisoara, and Pitesti, have proposed future projects on autonomous and connected vehicles.

11.4.1 *General opinion, barriers to the implementation of AVs in urban areas and measures to overcome them*

The first part of the interview captures the general perception of the interviewees on the AVs topic. Almost unanimously, experts agree on a number of advantages, such as the reduction in the number of traffic accidents, minimization of human effort, the potential of social inclusion, more efficient use of travel time, improvement of urban mobility, and higher travel comfort and a reduction of transport externalities and negative impacts, such as pollution, which is fully in line with the findings in the international literature (Botello et al., 2019; Kamruzzaman et al., 2015). Experts agree that streets will have to change and develop if AVs will become a dominant form of transport. For example, road administration might consider if some safety features, such as raised kerbs, are beneficial. On the other hand, a distracted human driver might accidentally cross over the curb or into another lane, while the likelihood of such accidents is expected to be lower with AVs. Some respondents also emphasized the reduction and therefore a traffic flow improvement, more efficient transport logistics, a decrease in the number of vehicles on the street, lower vehicle emissions, and improved economic efficiency.

When asked about key barriers for the implementation of ACT, most experts identified (i) high purchase cost of the vehicles; (ii) hesitation of people to adopt the new technologies due to concerns about safety, privacy, and job loss; (iii) need for significant investment in infrastructure along with technical challenges related to ensuring of reliable communication between vehicles and infrastructure; and (iv) uncertain establishment of legal liability for accidents involving AVs, similar to the ones expressed in the literature on the topic (Chen et al., 2022; Holthausen et al., 2022; Kellerman, 2018; Lee & Hess, 2022). Furthermore, the majority of experts considered that an obstacle to the implementation of autonomous public transport/SAVs in several Romanian cities is the absence of segregated lanes for public transport.

Some of the mitigation measures proposed by the experts included a better understanding of disruptive factors and the necessity of coordination efforts between decision-makers and all the stakeholders involved. Information, communication campaigns, and pilot demonstrations were also mentioned for better understanding of the ACT benefits and mitigating the negative perceptions about safety (travelling in the AVs and on-street interaction with AVs), security, and costs. At an individual level, the experts mentioned the necessity of funding, tax exemptions, and other incentives for companies that invest in AVs infrastructure.

11.4.2   *Type of transport better suited for AVs use*

Interviewed experts agreed that if proven reliable, the AVs could contribute to a reduction in car ownership rates due to car sharing, as suggested also by other authors (Fagnant & Kockelman, 2014). Experts consider that the existing autonomous public transport (metro, trains, and trams) could be a good intermediary phase in the transition toward a fully public transport automation. Many of them envisaged a major optimization in operational efficiency of public transport due to the 24/7 operation of the AVs. Experts foresee other economic benefits with AVs being the last/first mile transport means for accessing the main public transport network, transport on demand, and logistics. A couple of experts pointed out that AVs would work better in a system of connected and AVs.

Overall, the experts recommended that current and future AVs-based shared mobility should be better explained to the Romanian audience in an effort to change the citizens' mobility habits and start developing a culture of smart and sustainable mobility.

11.4.3   *Pros and cons on AVs impact on urban structure*

The interviews also provided some insight into the relation between AVs and urban structure. It was suggested that AVs will lead to a better organization of road spaces that will have a positive impact on traffic flow as is also evidenced in the literature (Foster, 2017). Another pro is the foreseen reduction of the width of traffic lanes due to the AVs design, allowing more pedestrian/green space/micro-mobility-dedicated space, which is pointed out also in the international research (Schlossberg et al., 2018). Conversion of the current fuel stations into car parking spaces and refueling stations for AVs in the future, which is also consistent with the literature (Krueger et al., 2016), and the use of AVs to reduce the need for parking spaces, which is in total agreement with published work (Yigitcanlar et al., 2019), were seen as pros of AVs introduction.

Some of the experts highlighted the cons of AV introduction, for example, saying that the poor quality of infrastructure can lead to errors in driving. They also mentioned the Romanian cities' historically low capacity of road maintenance as an obstacle for operating the AVs transport. Experts suggested that local authorities should already start considering the implementation of intelligent transport systems (ITS), thus allowing for needed changes that will affect infrastructure and the electricity supply network while also considering the integration of renewable energy. The initial investment could be very high, but the benefits will be visible in the long term.

Respondents also highlighted the idea that AVs will influence the spatial aspects of the city, as urban sprawl could be a major challenge to introducing AVs in urban areas, given their accessibility and the utility of time spent in the vehicle (Anderson et al., 2014; Stead & Vaddadi, 2019). Moreover, experts expressed concerns regarding the increase of AVs private ownership which could potentially negatively impact congestion, and which may be specifically addressed through coherent policy interventions as indicated by other authors (Cohen & Cavoli, 2019).

It was also suggested that after AVs introduction, the sidewalk space, serving as parking space in most Romanian cities, could be used more efficiently, since vehicle parking will be better regulated and managed. Further, positive impact of curbside management will allow people to boarding/alighting autonomous public transport and shared vehicles, or loading/unloading goods into AVs. Experts have mentioned that the design of urban spaces and road infrastructure needs to consider the travel conditions for all types of vehicles during the transition to full AVs, including large vehicles – for emergency interventions, maximum speed limits, junction configuration, and turning possibilities.

A key idea was that the AVs have the potential to reduce the necessary parking spaces and the carbon footprint that is in line to opinions issued by other authors (Martinez & Viegas, 2017). Some of the interviewees also believe that, due to driverless cars, demand for parking around large shopping centers, often located on the outskirts of cities, will decrease, allowing for these centers to be replaced by mixed-use developments, which may enrich the city life. Another positive impact should be on the development and utilization of parking spaces in direct correlation with the vehicle fleet, residential and economic areas, travel needs, location of major travel attractions, etc. Optimization of park spaces is possible, according to experts, especially if authorities develop integrated databases. Overall, parking management depends more on the capacity of local authority to develop smart applications to inform vehicle users in real time about the availability of parking spaces and the possibility to book them when planning the travel.

The urban planning experts pointed out that in the early 2000s, in most Romanian cities, a number of suburban neighborhoods were built on the outskirts, but without taking into consideration the need of the inhabitants to move around and have access to the public transport network. This has been due to a lack of adequate planning legislation. In most of these neighborhoods, there is still no public transport because the road network is not designed for the circulation of public transport vehicles. This has led to dependence on the private car and a discontinuity in connecting these neighborhoods with the rest of the city. The proper correlation of urban planning regulations and autonomous and connected transport legislation has a high potential to ensure a better integration of these unconnected city neighborhoods. The experts stressed that ICT, through the intelligent use of public transport vehicles, can play a significant role in overcoming this type of fracture in the city structure, improving mobility and social inclusion.

One major benefit of AVs as seen by respondents is the reallocation of the parking spaces to new utilizations such as urban green spaces, pedestrian areas, bicycle parking spaces (preferably shared), or even new bicycle lanes, idea which is consistent with international literature on this topic (Silva et al., 2021).

11.4.4   Implications of AVs adoption for the urban and transport policy framework in Romania

There is little doubt among the experts that AVs will have a significant impact on urban and transport policies in the not-too-distant future. Public policies at central, regional, and even local levels will have to adapt to the new infrastructure needs and the new type of digital-only management. This aspect was mentioned in various forms by the experts interviewed. The autonomous transport shall be integrated and harmonized with all public services included in the smart city strategies. Local decision-makers and communities should start thinking of a shared digital platform for the common delivery of public services.

Experts expressed the idea that future developments will lead to local authorities losing some degree of decision – making autonomy and control over the infrastructure, as the decisions will be more data driven and faster. The potential for using AVs as data-source for planning was suggested as well, consistent with literature (Gavanas, 2019). Experts pointed that all strategic documents at local level (SUMPs and smart cities strategies) will require significant updates and improvements, a costly and time-consuming process, while digitalization could improve strategic planning. Additionally, autonomous transport, and in particularly, shared and public transport, could be a supporting component of sustainable mobility solutions if better integrated into the GUPs and sustainable mobility plans.

Most respondents foresee the emergence of new digital businesses in the transport sector and the mobility market. This is in line also with the findings in Cassetta et al. (2017). The mobility-as-a-service (MaaS) concept is still in its infancy in Romania and is currently insufficiently understood, but a proper MaaS business model could lead to a more sustainable mobility with reduced operating costs. Cooperation with the private sector is needed to reduce the costs of developing new transport and mobility services, and their dedicated infrastructure, whether physical or digital. These findings are similar to those found in available research (Ackermann, 2021; Ho et al., 2018; Sarasini et al., 2017).

11.4.5 Implications of AVs implementation on Romanian's urban planning legislations

The experts interviewed expressed the opinion that urban and transport planning can play a key role in the transition to AVs, due to its positive impacts, reducing their negative impacts and exploiting opportunities. The process will be more efficient if key priority areas are adopted at European and global levels with the aim of achieving standardization and acceptability on a large scale. This could enable a more rapid implementation of the AVs at the national level. Standardization could be achieved through public investments, while best practices are needed to strengthen the enabling framework for the development and deployment of AVs. Specific requirements should be explored to support industry to navigate more effectively the existing and future standards landscape. In particular, communications standards are quite complex, and users need guidance on how to use them. Also, further standardization is needed on infrastructure development projects. Standardization for digital infrastructure and data communication needs to be tackled in future regulations.

There is a consensus among the interviewed experts that autonomous transportation needs to be integrated with all forms of urban planning, focusing on society and community rather than solely focusing on the road traffic network. For experts, planning will become much more flexible, with fewer restrictions, and of paramount importance in diverting urban planning from solving the traffic problems currently generated, toward the development of the specific social and environmental needs of the community.

Most of the experts suggested that, in the Romanian case, the setbacks of the infrastructure, civic mentality, and education cannot be bridged by law. A lengthy process is therefore required in which all the above-specified problems can be addressed seriously and consistently. A key topic is to set up the regulatory framework by amending legislation to link urban development with the adoption of autonomous transport. For instance, amendments are needed to the law on spatial planning and urban development (Parlamentul Romaniei, 2001, p. 350) and the traffic road code (Guvernul Romaniei, 2006) is mandatory. At the European level, there is a certain regulatory support provided by the amendment of EU Regulation 2019/2144, as revised by the Union (2022a). Experts recommend the improvement of the legal/regulatory/policy context for planning and implementing innovative ACT-related business models and projects, including the implementation of new types of infrastructure for ACT ecosystems.

11.5 Conclusions

In this chapter, we have attempted to explore the impact that AVs deployment could have on urban development in Romania through (i) the analysis of specific national and local strategic documents and (ii) the opinions collected from experts on the relationship between AVs and urban structure. Few national-level strategic documents envisage the introduction/deployment of ACT and AVs, while the analysis of local-level strategic documents from 25 selected Romanian cities barely mentions – in only 14 cities – smart mobility projects. The conclusion of first research objective of the chapter is that, in this field, Romania is at a very early stage. The second and third objectives are addressed with results showing a significant convergence between Romanian experts opinions and findings in the international literature: the experts indicated that the implementation of AVs could be beneficial in Romania if they are used for public or shared transport; simultaneously, AVs can have a positive impact on urban space by reducing traffic flows, lane widths, parking, and gas stations space; barriers of future AVs implementation in Romania can be mitigated through information campaign for the citizens, a better understanding of disruptive factors and constructive collaboration between all stakeholders; from the urban and transport policy and legislation perspectives, the experts suggested that new autonomous transport shall be integrated and harmonized with all public services within the smart city strategies and that legislation needs to be updated in an effort to recognize the impact of AVs. However, given their accessibility and the utility of time spent in the vehicle, the influence of AVs on the spatial aspects of the city needs to be addressed through well-established urban policies as urban sprawl could be a major challenge to introducing AVs on streets in urban areas. Some urban planners have remarked that the smart use of AVs may play a significant role in bypassing the fracture in the city fabric, improving the mobility and social inclusion.

Limitations of this study relate to the fact that knowledge on AVs is scarce, thus more interviews are needed to increase the accuracy of the findings. In addition, the interviews do not refer to a specific urban area, thus the opinions are conditioned by the state of urban development in Romania. Future research is needed to confirm and deepen the results obtained in this study in an effort to develop the agenda for planning Romanian cities in the AVs era.

Acknowledgments

The authors are grateful to the Romanian experts in mobility and transport who shared with us their views on urban challenges associated with ACT.

References

Ackermann, M. (2021). Forms of MaaS. In M. Ackermann (Ed.), *Mobility-as-a-Service: The Convergence of Automotive and Mobility Industries* (pp. 65–89). Springer International Publishing. https://doi.org/10.1007/978-3-030-75590-4_4

Agriesti, S., Brevi, F., Gandini, P., Marchionni, G., Parmar, R., Ponti, M., & Studer, L. (2020). Impact of driverless vehicles on urban environment and future mobility. *Transportation Research Procedia, 49,* 44–59. https://doi.org/10.1016/j.trpro.2020.09.005

Almeaibed, S., Al-Rubaye, S., Tsourdos, A., & Avdelidis, N. P. (2021). Digital twin analysis to promote safety and security in autonomous vehicles. *IEEE Communications Standards Magazine, 5*(1), 40–46. https://doi.org/10.1109/MCOMSTD.011.2100004

Alonso-Mora, J., Samaranayake, S., Wallar, A., Frazzoli, E., & Rus, D. (2017). On-demand high-capacity ride-sharing via dynamic trip-vehicle assignment. *Proceedings of the National Academy of Sciences*, *114*(3), 462–467. https://doi.org/10.1073/pnas.1611675114

An, S., Nam, D., & Jayakrishnan, R. (2019). Impacts of integrating shared autonomous vehicles into a Peer-to-Peer ridesharing system. *Procedia Computer Science*, *151*, 511–518. https://doi.org/10.1016/j.procs.2019.04.069

Anderson, J. M., Nidhi, K., Stanley, K. D., Sorensen, P., Samaras, C., & Oluwatola, O. A. (2014). *Autonomous Vehicle Technology: A Guide for Policymakers*. Rand Corporation.

Andorka, S., & Rambow-Hoeschele, K. (2020). Ethical and Social Aspects of Connected and Autonomous Vehicles: A Focus on Stakeholders' Responsibility and Customers' Willingness to Share Data. In R. José, K. Van Laerhoven, & H. Rodrigues (Eds.), *3rd EAI International Conference on IoT in Urban Space* (pp. 17–22). Springer International Publishing. https://doi.org/10.1007/978-3-030-28925-6_2

Andrei, L., & Luca, O. (2022). Towards a sustainable mobility development in Romanian Cities. A comparative analysis of the sustainable urban mobility plans at the National level. *Management Ressearch and Practice*, *14*(1), 11. http://mrp.ase.ro/no141/f3.pdf

Andrei, L., Luca, O., & Gaman, F. (2022). Insights from user preferences on automated vehicles: Influence of socio-demographic factors on value of time in Romania case. *Sustainability*, *14*(17), Article 17. https://doi.org/10.3390/su141710828

Andrei, L., Negulescu, M. H., & Luca, O. (2022). Premises for the future deployment of automated and connected transport in Romania considering citizens' perceptions and attitudes towards automated vehicles. *Energies*, *15*(5), Article 5. https://doi.org/10.3390/en15051698

Arfiansyah, D., & Han, H. (2021). Bandung Smart City: The Digital Revolution for a Sustainable Future. In J. C. Augusto (Ed.), *Handbook of Smart Cities* (pp. 439–465). Springer International Publishing. https://doi.org/10.1007/978-3-030-69698-6_92

Backhaus, W., Rupprecht, S., & France, D. (2019). *Practitioner Briefing: Road Vehicle Automation in Sustainable Urban Mobility Planning*. Rupprecht Consult - Forschung & Beratung GmbH. https://www.h2020-coexist.eu/wp-content/uploads/2019/06/Automation-SUMP-Practitioner-Briefing-2019.pdf

Barabás, I., Todoruţ, A., Cordoş, N., & Molea, A. (2017). Current challenges in autonomous driving. *IOP Conference Series: Materials Science and Engineering*, *252*, 012096. https://doi.org/10.1088/1757-899X/252/1/012096

Barbour, N., Menon, N., Zhang, Y., & Mannering, F. (2019). Shared automated vehicles: A statistical analysis of consumer use likelihoods and concerns. *Transport Policy*, *80*, 86–93. https://doi.org/10.1016/j.tranpol.2019.05.013

Bennett, R., Vijaygopal, R., & Kottasz, R. (2019). Attitudes towards autonomous vehicles among people with physical disabilities. *Transportation Research Part A: Policy and Practice*, *127*, 1–17. https://doi.org/10.1016/j.tra.2019.07.002

Bezai, N., Medjdoub, B., Fadli, F., Chalal, M., & Al-Habaibeh, A. (2020, November). Autonomous Vehicles and Smart Cities – Future Directions of Ownership vs Shared Mobility. *56th ISOCARP Virtual World Planning Congress: Post-Oil City, Planning for Urban Green Deals*, Doha, Qatar. http://irep.ntu.ac.uk/id/eprint/42031/

Bissell, D., Birtchnell, T., Elliott, A., & Hsu, E. L. (2020). Autonomous automobilities: The social impacts of driverless vehicles. *Current Sociology*, *68*(1), 116–134. https://doi.org/10.1177/0011392118816743

BMDV. (2022). *Autonome-Fahrzeuge-Genehmigungs-und-Betriebs-Verordnung (AFGBV)*. Bundeskabinett verabschiedet Verordnung zum Autonomen Fahren. https://bmdv.bund.de/SharedDocs/DE/Pressemitteilungen/2022/008-wissing-verordnung-zum-autonomen-fahren.html

Bösch, P. M., Becker, F., Becker, H., & Axhausen, K. W. (2021). How will Autonomous Vehicles Impact Car Ownership and Travel Behavior. In R. Vickerman (Ed.), *International Encyclopedia of Transportation* (pp. 508–513). Elsevier. https://doi.org/10.1016/B978-0-08-102671-7.10093-4

Botello, B., Buehler, R., Hankey, S., Mondschein, A., & Jiang, Z. (2019). Planning for walking and cycling in an autonomous-vehicle future. *Transportation Research Interdisciplinary Perspectives*, *1*, 100012. https://doi.org/10.1016/j.trip.2019.100012

Campisi, T., Severino, A., Al-Rashid, M. A., & Pau, G. (2021). The development of the smart cities in the connected and autonomous vehicles (CAVs) era: From mobility patterns to scaling in cities. *Infrastructures*, *6*(7), Article 7. https://doi.org/10.3390/infrastructures6070100

Carabulea, L., Pozna, C., Antonya, C., Husar, C., & Băicoianu, A. (2022). The influence of the advanced emergency braking system in critical scenarios for autonomous vehicles. *IOP Conference Series: Materials Science and Engineering*, *1220*(1), 012045. https://doi.org/10.1088/1757-899X/1220/1/012045

Cassandras, C. G. (2017). Automating mobility in smart cities. *Annual Reviews in Control*, *44*, 1–8. https://doi.org/10.1016/j.arcontrol.2017.10.001

Cassetta, E., Marra, A., Pozzi, C., & Antonelli, P. (2017). Emerging technological trajectories and new mobility solutions. A large-scale investigation on transport-related innovative start-ups and implications for policy. *Transportation Research Part A: Policy and Practice*, *106*, 1–11. https://doi.org/10.1016/j.tra.2017.09.009

Cavoli, C., Phillips, B., Cohen, T., & Jones, P. (2017). *Social and Behavioural Questions Associated with Automated Vehicles: A Literature Review*. https://trid.trb.org/view/1457834

Chen, Y., Shiwakoti, N., Stasinopoulos, P., & Khan, S. K. (2022). State-of-the-art of factors affecting the adoption of automated vehicles. *Sustainability*, *14*(11), Article 11. https://doi.org/10.3390/su141116697

Clewlow, R. R., & Mishra, G. S. (2017). *Disruptive Transportation: The Adoption, Utilization, and Impacts of Ride-Hailing in the United States*. https://escholarship.org/uc/item/82w2z91j

Cohen, T., & Cavoli, C. (2019). Automated vehicles: Exploring possible consequences of government (non)intervention for congestion and accessibility. *Transport Reviews*, *39*(1), 129–151. https://doi.org/10.1080/01441647.2018.1524401

Correia, G. H. de A., Looff, E., van Cranenburgh, S., Snelder, M., & van Arem, B. (2019). On the impact of vehicle automation on the value of travel time while performing work and leisure activities in a car: Theoretical insights and results from a stated preference survey. *Transportation Research Part A: Policy and Practice*, *119*, 359–382. https://doi.org/10.1016/j.tra.2018.11.016

Crute, J., Riggs, W., Chapin, T. S., & Stevens, L. (2018). *Planning for Autonomous Mobility*. American Planning Association.

De Lara, S. (2020). *The Driverless City: How Will AVs Shape Cities in the Future?* https://doi.org/10.13140/RG.2.2.10602.95689

Dennis, E. P., Spulber, A., Sathe Brugerman, V., Kuntzsch, R., & Neuner, R. (2017). *Planning for Connected and Automated Vehicles*. Technology Research. https://www.cargroup.org/publication/planning-for-connected-and-automated-vehicles/

Duvall, T., Hannon, E., Katseff, J., Safran, B., & Wallace, T. (2019). *A new look at autonomous-vehicle infrastructure | McKinsey* (What infrastructure improvements will promote the growth of autono-mous vehicles while simultaneously encouraging shared ridership?; p. 6). McKinsey & Company. https://www.mckinsey.com/industries/travel-logistics-and-infrastructure/our-insights/a-new-look-at-autonomous-vehicle-infrastructure

Eppenberger, N., & Richter, M. A. (2021). The opportunity of shared autonomous vehicles to improve spatial equity in accessibility and socio-economic developments in European urban areas. *European Transport Research Review*, *13*(1), 32. https://doi.org/10.1186/s12544-021-00484-4

Fagnant, D. J., & Kockelman, K. M. (2014). The travel and environmental implications of shared autonomous vehicles, using agent-based model scenarios. *Transportation Research Part C: Emerging Technologies*, *40*, 1–13. https://doi.org/10.1016/j.trc.2013.12.001

FMTDI. (2015). *Strategy for Automated and Connected Driving*. Federal Ministry of Transport and Digital Infrastructure, Germany. https://bmdv.bund.de/SharedDocs/EN/publications/strategy-for-automated-and-connected-driving.pdf?__blob=publicationFile

Foster, R. (2017, July 11). Integrating autonomous vehicles into complete streets. *Medium*. https://medium.com/@robert.m.fostr/integrating-autonomous-vehicles-into-complete-streets-e7f930c150b5

Fraedrich, E., Heinrichs, D., Bahamonde-Birke, F. J., & Cyganski, R. (2019). Autonomous driving, the built environment and policy implications. *Transportation Research Part A: Policy and Practice*, *122*, 162–172. https://doi.org/10.1016/j.tra.2018.02.018

Galich, A., & Stark, K. (2021). How will the introduction of automated vehicles impact private car ownership? *Case Studies on Transport Policy*, *9*(2), 578–589. https://doi.org/10.1016/j.cstp.2021.02.012

Gavanas, N. (2019). Autonomous road vehicles: Challenges for urban planning in European cities. *Urban Science*, *3*(2), Article 2. https://doi.org/10.3390/urbansci3020061

González-González, E., Nogués, S., & Stead, D. (2020). Parking futures: Preparing European cities for the advent of automated vehicles. *Land Use Policy*, *91*, 104010. https://doi.org/10.1016/j.landusepol.2019.05.029

Guvernul Romaniei, I. (2006). *Ordonanţa de urgenţă nr. 195/2002 privind circulaţia pe drumurile publice actualizat 2022*. Guvernul Romaniei. https://lege5.ro/Gratuit/heztgmjx/ordonanta-de-urgenta-nr-195-2002-privind-circulatia-pe-drumurile-publice

Gyergyay, B., Gomari, S., Friedrich, M., Sonnleitner, J., Olstam, J., & Johansson, F. (2019). Automation-ready framework for urban transport and road infrastructure planning. *Transportation Research Procedia*, *41*, 88–97. https://doi.org/10.1016/j.trpro.2019.09.018

Hamadneh, J., & Esztergár-Kiss, D. (2019). Impacts of Shared Autonomous Vehicles on the Travelers' Mobility. *2019 6th International Conference on Models and Technologies for Intelligent Transportation Systems (MT-ITS)*, 1–9. https://doi.org/10.1109/MTITS.2019.8883392

Harper, C. D., Hendrickson, C. T., Mangones, S., & Samaras, C. (2016). Estimating potential increases in travel with autonomous vehicles for the non-driving, elderly and people with travel-restrictive medical conditions. *Transportation Research Part C: Emerging Technologies*, *72*, 1–9. https://doi.org/10.1016/j.trc.2016.09.003

Hinderer, H., Stegmüller, J., Schmidt, J., Sommer, J., & Lucke, J. (2018). Acceptance of Autonomous Vehicles in Suburban Public Transport. *2018 IEEE International Conference on Engineering, Technology and Innovation (ICE/ITMC)*, 1–8. https://doi.org/10.1109/ICE.2018.8436261

Ho, C. Q., Hensher, D. A., Mulley, C., & Wong, Y. Z. (2018). Potential uptake and willingness-to-pay for Mobility as a Service (MaaS): A stated choice study. *Transportation Research Part A: Policy and Practice*, *117*, 302–318. https://doi.org/10.1016/j.tra.2018.08.025

Holthausen, B. E., Stuck, R. E., & Walker, B. N. (2022). Trust in Automated Vehicles. In A. Riener, M. Jeon, & I. Alvarez (Eds.), *User Experience Design in the Era of Automated Driving* (pp. 29–49). Springer International Publishing. https://doi.org/10.1007/978-3-030-77726-5_2

Hunter, C. B. (2021). *Fleet operations, curb usage, and parking search for shared autonomous vehicle (SAV) fleets* [Thesis]. https://doi.org/10.26153/tsw/43229

Hwang, J., Li, W., Stough, L., Lee, C., & Turnbull, K. (2020). A focus group study on the potential of autonomous vehicles as a viable transportation option: Perspectives from people with disabilities and public transit agencies. *Transportation Research Part F: Traffic Psychology and Behaviour*, *70*, 260–274. https://doi.org/10.1016/j.trf.2020.03.007

IIHS-HLDI. (2023). *Advanced driver assistance: Autonomous vehicle laws.* IIHS-HLDI Crash Testing and Highway Safety. https://www.iihs.org/topics/advanced-driver-assistance/autonomous-vehicle-laws

Ionescu, C. A., Fülöp, M. T., Topor, D. I., Căpușneanu, S., Breaz, T. O., Stănescu, S. G., & Coman, M. D. (2021). The new era of business digitization through the implementation of 5G technology in Romania. *Sustainability*, *13*(23), Article 23. https://doi.org/10.3390/su132313401

Johnson, E., & Nica, E. (2021). Connected vehicle technologies, autonomous driving perception algorithms, and smart sustainable urban mobility behaviors in networked transport systems. *Contemporary Readings in Law and Social Justice*, *13*, 37. https://heinonline.org/HOL/Page?handle=hein.journals/conreadlsj13&id=131&div=&collection=

Kamruzzaman, M., Hine, J., & Yigitcanlar, T. (2015). Investigating the link between carbon dioxide emissions and transport-related social exclusion in rural Northern Ireland. *International Journal of Environmental Science and Technology*, *12*(11), 3463–3478. https://doi.org/10.1007/s13762-015-0771-8

Kane, M., & Whitehead, J. (2017). How to ride transport disruption – A sustainable framework for future urban mobility. *Australian Planner*, *54*(3), 177–185. https://doi.org/10.1080/07293682.2018.1424002

Kassens-Noor, E., Cai, M., Kotval-Karamchandani, Z., & Decaminada, T. (2021). Autonomous vehicles and mobility for people with special needs. *Transportation Research Part A: Policy and Practice*, *150*, 385–397. https://doi.org/10.1016/j.tra.2021.06.014

Kellerman, A. (2018). *Automated and Autonomous Spatial Mobilities.* Edward Elgar Publishing Ltd. https://www.e-elgar.com/shop/gbp/automated-and-autonomous-spatial-mobilities-9781786438485.html

Kester, J. (2022). Insuring future automobility: A qualitative discussion of British and Dutch car insurer's responses to connected and automated vehicles. *Research in Transportation Business & Management*, *45*, 100903. https://doi.org/10.1016/j.rtbm.2022.100903

Kisner, C., Fillin-Yeh, K., Bharadwaj, S., Schmidt, C., Abdulsamad, M., & Engel, A. (2019). *Blueprint for Autonomous Urbanism Second edition.* NACTO. https://nacto.org/publication/bau2/transit/

Krueger, R., Rashidi, T. H., & Rose, J. M. (2016). Preferences for shared autonomous vehicles. *Transportation Research Part C: Emerging Technologies*, *69*, 343–355. https://doi.org/10.1016/j.trc.2016.06.015

Lee, D., & Hess, D. J. (2022). Public concerns and connected and automated vehicles: Safety, privacy, and data security. *Humanities and Social Sciences Communications*, *9*(1), Article 1. https://doi.org/10.1057/s41599-022-01110-x

Lowenberg-DeBoer, J., Behrendt, K., Ehlers, M.-H., Dillon, C., Gabriel, A., Huang, I. Y., Kumwenda, I., Mark, T., Meyer-Aurich, A., Milics, G., Olagunju, K. O., Pedersen, S. M., Shockley, J., & Rose, D. (2022). Lessons to be learned in adoption of autonomous equipment for field crops. *Applied Economic Perspectives and Policy*, *44*(2), 848–864. https://doi.org/10.1002/aepp.13177

Luca, O., Gaman, F., & Răuță, E. (2021). Towards a national harmonized framework for urban plans and strategies in Romania. *Sustainability*, *13*(4), Article 4. https://doi.org/10.3390/su13041930

Luu, D. L., Lupu, C., & Chirita, D. (2019). Design and Development of Smart Cars Model for Autonomous Vehicles in a Platooning. *2019 15th International Conference on Engineering of Modern Electric Systems (EMES)*, 21–24. https://doi.org/10.1109/EMES.2019.8795199

Martinez, L. M., & Viegas, J. M. (2017). Assessing the impacts of deploying a shared self-driving urban mobility system: An agent-based model applied to the city of Lisbon, Portugal. *International Journal of Transportation Science and Technology*, *6*(1), 13–27. https://doi.org/10.1016/j.ijtst.2017.05.005

Medina-Tapia, M., & Robusté, F. (2019). Implementation of connected and autonomous vehicles in cities could have neutral effects on the total travel time costs: Modeling and analysis for a circular city. *Sustainability*, *11*(2), Article 2. https://doi.org/10.3390/su11020482

Millard-Ball, A. (2018). Pedestrians, autonomous vehicles, and cities. *Journal of Planning Education and Research*, *38*(1), 6–12. https://doi.org/10.1177/0739456X16675674

Millard-Ball, A. (2019). The autonomous vehicle parking problem. *Transport Policy*, *75*, 99–108. https://doi.org/10.1016/j.tranpol.2019.01.003

Mo, B., Cao, Z., Zhang, H., Shen, Y., & Zhao, J. (2021). Competition between shared autonomous vehicles and public transit: A case study in Singapore. *Transportation Research Part C: Emerging Technologies*, *127*, 103058. https://doi.org/10.1016/j.trc.2021.103058

MTI. (2022). *Strategia Națională STI - 2020-2030*. Ministerul Transporturilor si Infrastructurii. https://www.mt.ro/web14/documente/domenii/Sisteme-de-transport-inteligente/Acte-normative/2.pdf

Nastjuk, I., Herrenkind, B., Marrone, M., Brendel, A. B., & Kolbe, L. M. (2020). What drives the acceptance of autonomous driving? An investigation of acceptance factors from an end-user's perspective. *Technological Forecasting and Social Change*, *161*, 120319. https://doi.org/10.1016/j.techfore.2020.120319

Naumov, S., Keith, D. R., & Fine, C. H. (2020). Unintended consequences of automated vehicles and pooling for urban transportation systems. *Production and Operations Management*, *29*(5), 1354–1371. https://doi.org/10.1111/poms.13166

NHTSA. (2016). *Federal Automated Vehicles Policy—Accelerating the Next Revolution in Roadway Safety*. US Department of Transportation. https://www.nhtsa.gov/sites/nhtsa.gov/files/federal_automated_vehicles_policy.pdf

NITPA. (2017). *Paths to a self driving future*. Ministry of Infrastructure and the Environment. https://knowledge-base.connectedautomateddriving.eu/wp-content/uploads/2019/08/Pathstoaself_drivingfuture.pdf

Nourinejad, M., Bahrami, S., & Roorda, M. J. (2018). Designing parking facilities for autonomous vehicles. *Transportation Research Part B: Methodological*, *109*, 110–127. https://doi.org/10.1016/j.trb.2017.12.017

OpenStreetMap. (2021, January 29). *OpenStreetMap*. https://www.openstreetmap.org/

Ostermeijer, F., Koster, H. R. A., & van Ommeren, J. (2019). Residential parking costs and car ownership: Implications for parking policy and automated vehicles. *Regional Science and Urban Economics*, *77*, 276–288. https://doi.org/10.1016/j.regsciurbeco.2019.05.005

Papa, E., & Ferreira, A. (2018). Sustainable Accessibility and the Implementation of Automated Vehicles: Identifying Critical Decisions. *Urban Science*, *2*(1), Article 1. https://doi.org/10.3390/urbansci2010005

Park, J. E., Byun, W., Kim, Y., Ahn, H., & Shin, D. K. (2021). The impact of automated vehicles on traffic flow and road capacity on urban road networks. *Journal of Advanced Transportation*, *2021*, e8404951. https://doi.org/10.1155/2021/8404951

Parlamentul Romaniei. (2001). *Legea nr. 350/2001 privind amenajarea teritoriului şi urbanismul actualizată 2022*. Parlamentul Romaniei. https://lege5.ro/Gratuit/gmztknju/legea-nr-350-2001-privind-amenajarea-teritoriului-si-urbanismul

Pauca, O., Maxim, A., & Caruntu, C.-F. (2021). Control architecture for cooperative autonomous vehicles driving in platoons at highway speeds. *IEEE Access*, *9*, 153472–153490. https://doi.org/10.1109/ACCESS.2021.3128235

Perișoară, L. A., Dănișor, C., & Săcăleanu, D. I. (2022). Analysis of Mobile Communications Services for Internet of Things in Romania. *2022 23rd International Carpathian Control Conference (ICCC)*, 198–202. https://doi.org/10.1109/ICCC54292.2022.9805810

Sarasini, S., Sochor, J., & Arby, H. (2017). *What Characterises a Sustainable MaaS Business Model?*. ICoMaaS 2017.1st international conference on Mobility as a Service Tampere 28–29.11.2017. https://urn.kb.se/resolve?urn=urn:nbn:se:ri:diva-33094

Schlossberg, M., Riggs, W., Millard-Ball, A., & Shay, E. (2018). *Rethinking the Street in an Era of Driverless Cars.* https://doi.org/10.13140/RG.2.2.29462.04162

Silva, D. S., Csiszár, C., & Földes, D. (2021). Autonomous vehicles and urban space management. *Zeszyty Naukowe. Transport/Politechnika Śląska, 110,* 169. https://bibliotekanauki.pl/articles/2116417

Stead, D., & Vaddadi, B. (2019). Automated vehicles and how they may affect urban form: A review of recent scenario studies. *Cities, 92,* 125–133. https://doi.org/10.1016/j.cities.2019.03.020

Trasnea, B., Marina, L. A., Vasilcoi, A., Pozna, C. R., & Grigorescu, S. M. (2019). GridSim: A Vehicle Kinematics Engine for Deep Neuroevolutionary Control in Autonomous Driving. *2019 Third IEEE International Conference on Robotic Computing (IRC),* 443–444. https://doi.org/10.1109/IRC.2019.00091

UK Statutory Instruments. (2022). *Automated and Electric Vehicles in Primary and Secondary Legislation.* Legislation.Gov.Uk. https://www.legislation.gov.uk/primary+secondary?title=Automated%20and%20Electric%20Vehicles

Union, P. O. of the E. (2020, December 9). *COM/2020/789 Final, Communication from the Commission to the European Parliament, the Council, the European Economic and Social Committee and the Committee of the Regions Sustainable and Smart Mobility Strategy – Putting European Transport on Track for the Future* [Website]. Publications Office of the EU; Publications Office of the European Union. https://op.europa.eu/en/publication-detail/-/publication/5e601657-3b06-11eb-b27b-01aa75ed71a1/language-en/format-PDF/source-286349461

Union, P. O. of the E. (2022a, June 8). *Commission Delegated Regulation (EU) /... Amending Regulation (EU) 2019/2144 of the European Parliament and of the Council to Take into Account Technical Progress and Regulatory Developments Concerning Amendments to Vehicle Regulations Adopted in the Context of the United Nations Economic Commission for Europe, C/2022/3610 final* [Website]. Publications Office of the European Union. http://op.europa.eu/en/publication-detail/-/publication/d3cc8851-e71c-11ec-a534-01aa75ed71a1/language-en/format-PDF

Union, P. O. of the E. (2022b, August 5). *C/2022/5402, Commission Implementing Regulation (EU) 2022/1426 of 5 August 2022 Laying Down Rules for the Application of Regulation (EU) 2019/2144 of the European Parliament and of the Council as Regards Uniform Procedures and Technical Specifications for the Type-Approval of the Automated Driving System (ADS) of Fully Automated Vehicles (Text with EEA relevance)* [Website]. Publications Office of the EU; Publications Office of the European Union. https://op.europa.eu/en/publication-detail/-/publication/94bfefa8-24e9-11ed-8fa0-01aa75ed71a1/language-en/format-PDF/source-286349790

Yigitcanlar, T., Wilson, M., & Kamruzzaman, M. (2019). Disruptive impacts of automated driving systems on the built environment and land use: An urban planner's perspective. *Journal of Open Innovation: Technology, Market, and Complexity, 5*(2), Article 2. https://doi.org/10.3390/joitmc5020024

Zhang, W., Guhathakurta, S., & Khalil, E. B. (2018). The impact of private autonomous vehicles on vehicle ownership and unoccupied VMT generation. *Transportation Research Part C: Emerging Technologies, 90,* 156–165. https://doi.org/10.1016/j.trc.2018.03.005

Zhou, Y., Sato, H., & Yamamoto, T. (2021). Shared low-speed autonomous vehicle system for suburban residential areas. *Sustainability, 13*(15), Article 15. https://doi.org/10.3390/su13158638

12 Public–private partnership (PPP) and ICT in a mega-smart city

The case of Istanbul

Sabina Klimek

12.1 Introduction

Public–private partnership (PPP) has become a dominant global model for public infrastructure and service provision in the 21st century (Willems et al., 2018). PPP initiatives create the opportunity for stakeholders to share risk among public and private sector actors, while at the same time enabling the public sector to benefit from the private sector know-how and experience (Malek and Gundaliya, 2021; Malik and Kaur, 2021). Increasingly, in the process of implementing information and communication technology (ICT)-based solutions, city authorities decided to introduce a model of PPP. This allowed them to implement larger, more expensive, and complex projects efficiently (Selim and ElGohary, 2020; Gobin-Rahimbux et al., 2020). The case of the city of Istanbul showcases how the utilization of the PPP model has proved consequential in the process of the city transitioning to becoming a smart city (cf. Visvizi and Lytras, 2019). In other words, the growing importance of Istanbul as a smart city would not have taken place without the PPP models introduced by the city mayor and the city's managers, thanks to whom innovative technologies are implemented in the city. This chapter will discuss these. Especially after 2015, a specific plan and road map have been determined. Istanbul is ranked top in the world smart city rankings every year (Smart City Index, 2022). The Istanbul Metropolitan Municipality recently played a prominent role in smart city studies via the Smart Cities Directorate and in the Smart City Istanbul Project. In addition, the Istanbul Metropolitan Municipality has established a company that is responsible for creating, buying, and introducing innovative technologies to push Istanbul to the forefront of smart cities in the world (Report ISBAK, 2023). The objective of this chapter is to present the new business models based on PPP in the future development trends of smart cities, connected with smart management and smartification of the quality of life based on the case of Istanbul. It is analyzed how the implemented PPP models influenced the development of the city of Istanbul as a smart city. The argument in this chapter is structured as follows. First, PPP opportunities for smart projects are presented. Next, the models that play a significant role in the implementation of PPP projects in the area of

DOI: 10.1201/9781003415930-16

smart infrastructure in Turkey are indicated. Subsequently, the solutions used by the city of Istanbul in PPP projects are introduced. Conclusions follow.

12.2   Public–private partnership and smart city projects

Global trends in social and economic development now include urbanization. The rate of urbanization throughout the world has dramatically risen in recent decades. Every week, more individuals are moving to cities (Liu *et al.*, 2020). Many governments have endorsed the creation of smart cities as a successful strategy to enhance urban development and management (Selim and ElGohary, 2020). The term "smart city" originates in marketing campaigns of international ICT companies, which perceived urban environments as a strategic interest, as a "huge untapped market", with annual investments of hundreds of billions of dollars (Singh and Singla, 2021). Historically, the development of cities, including their expansion and decline, has depended on a variety of factors, including the industry around which a city was established, the specialization that resulted, and the capacity of a specific city to adapt to shifting social, political, and historical conditions (Visvizi and Lytras, 2019). The first generation of smart city initiatives, this time led by city authorities, rather than the business sector, emerged around 2008 in the aftermath of the global financial crisis. Local public sector budget cuts and post-recession corporate marketing strategies in the private sector had profound influence on the narrative and the shape of smart city initiatives, in some ways representing a post-recession transformation strategy (Voorwinden, 2021).

Solutions provided to cities by entrepreneurs and investors, mainly from the ICT industry, were then referred to as the Smart City 1.0 model (Perez-delHoyo et al., 2019, Rathore et al., 2017). This term came without an understanding of the needs and expectations of other stakeholders, such as local government authorities or local communities (Jonek-Kowalska and Kaźmierczak, 2021). When Smart City 1.0 model became the subject of criticism, the idea of Smart City 2.0 was formulated, in which local authorities became the leaders of Smart City initiatives. For local governments, Smart City offers potential solutions in the context of intensive urbanization and economic savings (Perez-delHoyo et al., 2019; Voorwinden, 2021). Consecutively, the idea of Smart City 3.0 was presented—this time including the inhabitants as participants in the processes of developing innovative functions for urban spaces. However, the constantly deteriorating condition of the natural environment made it necessary to supplement the Smart City concept with ecological issues. In this way, the smart city became a sustainable city—Sustainable Smart City (Jonek-Kowalska and Kaźmierczak, 2021).

Research suggests smart city projects use cutting-edge technology to increase the efficiency of urban operations (Tan and Taeihagh, 2020). The creation of smart cities is fraught with several difficulties. Governments across the world struggle with fiscal restrictions and aging infrastructure (Lam and Yang, 2020). Restrictions on public funding impede the growth of smart city programs. Additionally, the development of smart cities involves a variety of parties, including citizens, municipal authorities, and nonprofit and private organizations with a range of objectives (Seema and John, 2016). Moreover, the construction of a smart city necessitates a high level of innovation, necessitating a lot of commercial and business ingenuity from service providers (Ojasalo and Kauppinen, 2016). To promote smart city projects, collaborations between business and nonprofit organizations are necessary.

PPP is one of the most frequently discussed topics in the area of investment activity for states, cities, and local governments (Willems *et al.*, 2018; Andon, 2012; Broadbent and Laughlin, 2004). In a slightly narrower sense, PPP is defined as a collaboration between the public and private sectors with the goal of carrying out initiatives or offering services typically supplied by the public sector (EuropeanCourt of Auditors, 2018). PPPs offer a cooperative framework for the public and private sectors to develop smart city initiatives (Liu *et al.*, 2020). PPPs are a common strategy for combining the power of the public and private sectors, allowing the former to provide public infrastructure while the latter benefits from their money, creativity, and knowledge (Pellegrino *et al.*, 2019).

Due to limited financial resources from the public budget, authorities more frequently decide to work with private entities in PPPs. This form of cooperation is accepted by both parties, given that the results are mutually beneficial (Malek and Gundaliya, 2021). PPP provides enterprises with guaranteed, long-term profits and gives public entities the opportunity to expand technical or communication infrastructure without major financial costs (Alhassani and Almarri, 2020). In addition, the private sector includes high-quality specialists, which significantly speeds up work in terms of logistics and formalities.

Arguably, the main goal of PPP is to improve efficiency, increase revenues, and improve infrastructure through privately funded investments (Kuyucak and Sengur, 2020; Lam and Yang, 2020; Malek and Gundaliya, 2021). In practice, PPP is widely implemented in several sectors, such as water supply, sewage disposal, energy supply and production, development of public places (stadiums, schools, hospitals, healthcare facilities, and cemeteries), bioenergy projects, housing, telecommunications, defense, transport infrastructure (including roads and airports), metro network, ports, logistics systems, and railway stations (Kuyucak and Sengur, 2020). Threats related to the implementation of PPP projects should also be considered. These include loss of control over the service delivery process, threats to the proper functioning of public administration, excessive increases in fees paid by service recipients, lack of competition, and—in relation—lower quality of services provided (Jayasena *et al.*, 2020; Almarri, 2022).

It is considered that the essential elements of a smart city are city, citizens, knowledge, intelligence, innovation, smart systems, infrastructure, and urban technologies (Jayasena *et al.*, 2020). Smart cities are based on five basic factors: power, infrastructure, financial sources, technology, and talented human resources (Celikyay, 2017).

It proves that typical components of smart city initiatives include a high degree of creativity and clever solutions, sustainability objectives, and technological cooperation with nearby universities or research facilities (Selim *et al.*, 2018). Therefore, traditional contract-based PPP models may not be appropriate for smart city initiatives (Liu *et al.*, 2020).

The precondition for smart cities is therefore the development of smart infrastructure. Given the shortcomings of traditional procurement and funding models, specific PPP models are required for information systems development (Jayasena et al., 2020). PPP can be used in infrastructure projects in the following ways (Almarri, 2022):

1 The Pan City model: providing smart solutions for existing cities by combining design and information technology
2 City modernization model: smart modernization of the existing built-up area and strengthening the current infrastructure to achieve the smart city goal
3 Model of urban redevelopment: replacing existing infrastructure with new infrastructure that meets the needs of the future.
4 Model of city development: identifying potential trends in city development and then supporting the current infrastructure in these directions.

From the above considerations, attention must first be drawn to the reversal of the relationship between business and local government; this includes emphasizing that the responsible party for reporting the need for intelligent solutions to city authorities should be local and regional authorities, not entrepreneurs and investors. Next, city dwellers are involved in the process of identifying urban needs and expectations; they should participate when decisions are being made regarding the conditions in which they live. Smart city is a holistic concept aimed at solving urbanization problems in cities; it requires significant financial outlays and technological innovations to improve the quality of life and well-being of residents, as well as to achieve sustainable social impact.

12.3   PPP models for delivering smart infrastructure in Turkey

It is important to contextualize the PPPs in Istanbul; the broader political and economic context of Turkey is key in this respect. Over the last ten years, we have seen a rapid increase in investment in Turkey, especially in infrastructure, hospitality, and tourism. All investments are strongly related to modern technological solutions and improving the quality of life, especially in cities. In 2021, the share of urban population in Turkey remained nearly unchanged at around 76.57% (Turkish Investment Office, 2022). Turkey has actively embraced PPP as a key approach to infrastructure development and service provision in various sectors. The country recognizes the potential of PPPs to leverage private sector expertise, resources, and efficiency in delivering public projects and services.

Over 2003–2016, PPP investments in Turkey amounted to nearly USD 150 billion, whereas the expected PPP investment for 2017–2023 is worth USD 325 billion in areas such as transportation, healthcare, and energy (Turkish Investment Office, 2022)). In Turkey, the use of ICT in environmental protection is becoming more common. As a developing country, Turkey is monitoring environmental health, gathering data, and working to improve cooperation with related international institutions. Linking ICT to environmental issues is a very new solution in developing countries—it is also new to Turkey (Gonel & Akinci, 2018). At the same time, smart buildings are becoming increasingly common in Turkey, because of the implementation of specific legislation. The majority of Turkey's smart buildings are increasingly common in big cities like Istanbul. Compared to conventional buildings, smart buildings use less energy, emit up to 40% less CO_2, use 40% less water, and produce 70% less water waste (The Ministry of Infrastructure Action Plan, 2013–2023). Turkey thus benefits from the implementation of a smart approach to construction in terms of energy savings. The transportation sector in Turkey is also benefiting from ICT solutions and—given the share of the automotive industry in the country—progress is substantially positive (Gonel & Akinci, 2018).

The Turkish economy is developing dynamically. Despite the current problems with inflation and weak currency, an average annual GDP growth rate is about 5.4% from 2002 to 2022 and it ranks 19th in economy in the world. The current investment trends of this country until 2040 are estimated to amount to USD 569 billion, with investment needs at USD 975 billion, creating an investment gap of USD 406 billion (Turkish Investment Office, 2022)). These data indicate that Turkey has huge investment needs. Probably, the country is not able to provide financing for all investments on its own; hence PPP seems to be a very good solution (Figures 12.1 and 12.2).

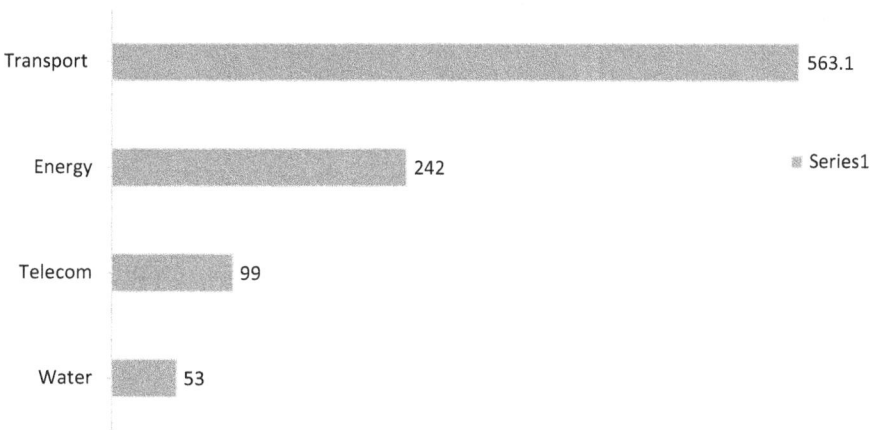

Figure 12.1 Investment needs by sectors (2022–2040).

Source: Author, based on: Turkish Investment Office (2022).

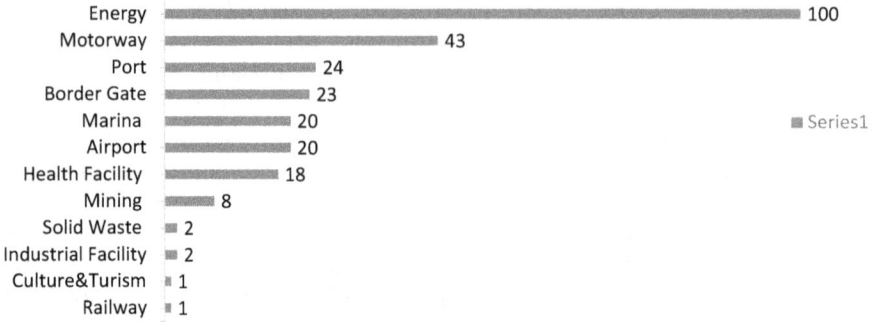

Figure 12.2 Breakdown of the PPP contracts (number).

Source: Author, based on: Turkish Investment Office (2022).

The value of PPP contracts in the chart displayed above amounts to USD 184 billion, with 86.5% of the amount falling into three main investment areas: airports (USD 89.640 billion), energy (USD 40.696 billion), and motorways (USD 28.876 billion).

Recent years of investment in Turkey have resulted in the country boasting a substantial number of new roads and highways, airports, and bridges. The implemented solutions boast state-of-the-art technological solutions, such as intelligent lighting, motorway entrance management, traffic monitoring, and bridge monitoring. At the same time, there is no shortage of criticism in the country that the investments on such a large scale were unnecessary and exaggerated. There are arguments that the country has many other needs, and these are not modern solutions in infrastructure. However, this does not change the fact that the above-mentioned investments are based on the latest ICT solutions, the implementation of which changes the quality of life of residents and improves their functioning.

Over 82% of funds allocated for PPP investments in transportation and communications in the years 2003–2021 relate to completed projects (Türkiye Investment Office, 2022). The main contract models on the Turkish PPP market in the years 1986 to 2021—in terms of both the value of investments and the number of projects—are Build–Operate–Transfer (BOT), Transfer of Operating Rights (TOR), Build–Lease and Transfer (BLT), and Build and Operate (BO).

As we can see in the Figures 12.3 and 12.4, the BOT model is the most frequently chosen model of cooperation between partners in Turkey, but not only. An arrangement where a facility is developed, funded, managed, and maintained by the concessionaire throughout the duration of the concession is known as a BOT, making it one of the more well-liked PPP choices. The concessionaire may or may not be the legal owner of the facility. The Channel Tunnel between the UK and France, Chinese power plants, Taiwanese high-speed rail, Kaohsiung MRT, and the Taipei Dome Complex are just a few

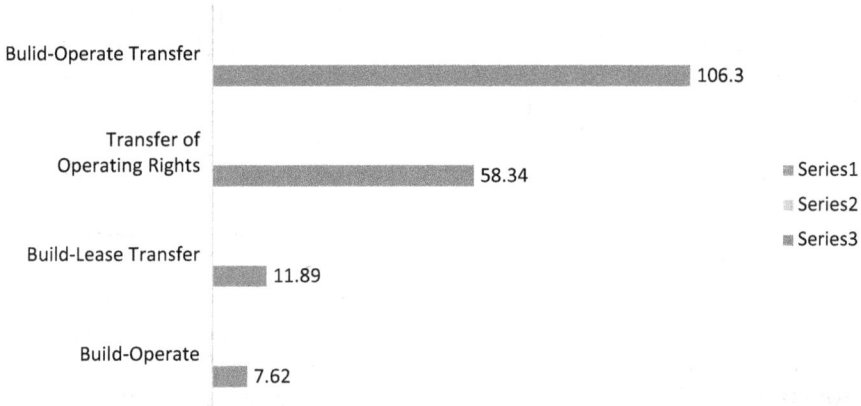

Figure 12.3 PPP Contract Models (USD Billion).
Source: Author, based on: Turkish Investment Office (2022).

Figure 12.4 PPP Contract Models (Number).
Source: Author, based on: Turkish Investment Office (2022).

examples of public infrastructure projects where a BOT was used to draw private participation (Chih-Yao and Ren-Jye, 2019). BOT models are considered the least risky for the government, but at the same time the most profitable for the private party.

Turkey has implemented many mega projects that are called smart infrastructure in the BOT model, e.g., it completed the 1915 Çanakkale Bridge project ($2.8 billion from BOT), Istanbul New Airport ($6.5 billion), and Eurasia Tunnel ($2 billion) (Türkiye Investment Office 2022: 30). In Turkey, PPP is also the primary method for financing and operating new airports and/or terminals. PPPs at Turkish airports date back to the mid-1990s, so the history spans more than 25 years. Since the first BOT project in the mid-1990s, two models have been used: BOT and Lease–Operate–Transfer (LOT). Antalya International Airport Terminal is a pioneer of BOT and PPP models at airports in Turkey. Due to the previously mentioned limitations, the operating rights of the terminal have been transferred to Antalya International Airport Terminal Management Inc. (AYT) in partnership with Bayindir Holding and Fraport for the last nine years. The terminal began operating on April 1, 1998. Istanbul Atatürk Airport, the second international terminal of Antalya Airport, was then launched. After the expiration of the BOT contract period and the transfer of Istanbul Ataturk Airport to SAA, the LOT model has appeared on the agenda in Turkey's aviation industry. The four main stages of BOT at an airport in Turkey are the planning and preparation phase, transaction phase, implementation phase, and transfer phase (Kuyucak and Sengur, 2020).

The above-mentioned projects in the BOT model are primarily characterized by a huge investment outlay, which would not be possible without the cooperation of the government and private parties. At the same time, these are projects that both bring a lot of elements from the area of smart city and smart investment and clearly influence the changes that are taking place both in Turkey and in the city they are concerned with, changing its structure, construction, and housing.

12.4   Istanbul and the PPP

As the most populous city in Turkey, Istanbul has significant investment potential. This flourishing metropolis is the cultural, commercial, and financial center of Turkey. Due to its size and functions, it is strategically positioned to play a significant role in the network of European cities. The metropolitan character of Istanbul is amplified by its innovative potential—undoubtedly the largest in the entire country. Strategically located at the crossroads of major roads to the sea, as well as the transportation intersection of Europe, Eurasia, and the Middle East, Istanbul has a population of over 15 million; the city has made great strides in recent years. However, it can be surprising that despite the innovative technology that is implemented in the area of transport, health, and energy, Istanbul went down in the Smart City Index. According to the fourth edition of the Smart City Index 2023 report, Istanbul is ranked 107th (out of 141), and in 2021 it was 88th out of 118 in the ranking of smart cities (Smart City Index, 2022). In the previous years, Istanbul was not taken into consideration in this research. As the most challenging problems in Istanbul, the authors of the report defined problems with air pollution, lack of green spaces, access to care sharing applications, traffic issues, and corruption. Also, housing was pointed out as the most important and problematic matter. Interestingly, issues related to the use of applications to solve everyday problems, such as information on traffic or arranging medical appointments, achieved the highest scores (Smart City Index, 2022). This may indicate the dependence that Istanbul manages to implement innovative technologies, but still has problems dealing with basic problems such as housing (Table 12.1).

It should be noted that the Istanbul city authorities are aware of the need to implement new technological solutions in Istanbul and are making every effort in this area. First of all, the city has set up its own internal organization

Table 12.1 Smart City Ranking 2023

Cities	2019	2020	2021	2023
Zurich	1	1	1	1
Oslo	2	2	2	2
Canberra				3
Copenhagen	4	3	5	4
Lausanne			4	5
London	3	10	3	6
Singapore	10	7	7	7
Helsinki	6	5	9	8
Geneva	7	8	6	9
Stockholm	9	9	11	10
Istanbul			88	107

Source: Author, based on Smart City Index, 2023.

responsible for the purchase, production, and implementation of smart solutions in Istanbul. The organization responsible for Istanbul's smart city projects is Istanbul Computing and Smart City Technologies Inc. (ISBAK). This organization's mission is to "develop sustainable Smart City projects based on next-generation technologies and improve the quality of life in cities, in particular Istanbul as one of Turkey's leading technology companies". Every year, ISBAK employees take part in several international events to gain knowledge and purchase solutions to improve functioning in Istanbul. At the same time, at least a dozen or so innovations in the area of the smart city are implemented (Report ISBAK, 2023).

Additionally, in order to achieve its goal of becoming a Focus City in 2023, the Special Commission for Smart Cities of the Istanbul Metropolitan Municipality was established in April 2015. The work of the commission was scheduled to last eight years, with the goal of implementing the listed actions supporting smart city applications (Çelikyay, 2017, Istanbul Metropolitan Municipality, 2023).

Information-based city applications and the development of intelligent transport systems are important works carried out by Istanbul in previous years. Istanbul's actions toward smart city applications are

- inclusion in a large consortium with 22 partners from all over Europe;
- inclusion in City SDK—Development of Toolkit and Smart City API for Service Developers;
- membership in the European Network of Living Labs (ENoLL) in order to complete the work necessary to create Smart City Living Labs;
- participation in many congresses and exhibitions abroad;
- support for Smart City Committees at Unites Cities Local Governments and regional organizations; and

- in 2021, Istanbul became the third Turkish city after Izmir and Ankara to join the European Bank for Reconstruction and Development Green Cities Program, which identifies, prioritizes, and connects cities' environmental challenges with sustainable infrastructure investments and policy measures.

(Çelikyay, 2017; European Bank for Reconstruction and Development (2021))

When we talk about Istanbul, we cannot fail to mention the largest smart investment in Turkey, which is Istanbul Grand Airport (IGA). It is the most strategic national mega project in Turkey, aiming to improve the capacity of existing airports. The Turkish government as well as the Istanbul Metropolitan Municipality collaborated on this mega project. In addition, the structural and architectural project was designed by nine international private companies, which are the Nordic Office of Architecture, Grimshaw, Arup Associates, Haptic Architects, Perkins+Will, Scott Brownrigg, Fonksiyon Mimarlık, TAM+Kiklop, and Pininfarina+Aecom. The bid was won by IGA, paying EUR 22.152 billion. The construction and development cost of the new Istanbul International Airport amounted to EUR 10.247 billion. The project was fully funded by the private sector. In total, 16-year loan agreements were signed between IGA and local/foreign banks, i.e., Ziraatbank, Halkbank, Vakifbank, Denizbank, Garantibank, and Finansbank. IGA has the right to operate the new airport for 25 years, which includes paying EUR 1.46 billion in rent to the Turkish government (Eren, 2018, Istanbul Metropolitan Municipality, 2023)).

The contribution of Istanbul's large airport to the national economy will amount to approximately USD 20 billion in 2025, corresponding to 4.9% of Turkey's gross national income. The airport will offer employment opportunities for 225,000 people. In addition, the organization of the mega project imposes an obligation on all enterprises operating at the airport to hold the leadership in energy and environmental design (LEED) green business certificates in order to make the Grand Airport in Istanbul the first green airport in the world (Eren, 2018; Delibasi, 2019).

The results of this study show that the new airport in Istanbul came to be due to top-level organizational management. Such an organizational structure

- reduces complexity, risk, and uncertainty in managing mega projects;
- improves the efficiency and quality of mega projects;
- reduces the risk associated with all actors in the organization;
- increases the harmony between the actors of mega projects; and
- increases the potential and capabilities of local institutions and companies.

(Eren, 2018; Delibasi, 2019)

In Turkey, there is a belief that the airport in Istanbul is a summary of Turkish skills in the field of construction, technology, modernity, and quality of services provided.

Summing up the issues of implementing smart solutions in Istanbul, it should be emphasized that every year, annual Istanbul PPP Week takes place that serves as an important event for the promotion of PPP. This meeting brings together influential infrastructure practitioners and strengthens collaboration between the public and private sectors. In addition, workshops covering topics such as the Environmental Focus Area are also held in Istanbul as part of the research on smart city projects (Istanbul PPP Week, 2023).

12.5   Conclusion

In Istanbul, 15 million people live, but in line with unofficial sources, it may be over 20 million people. It is the only city in the world that spans two continents. At the same time, it is a huge city with an area of 5,343 km² and access to three seas. These numbers allow us to understand what a huge metropolis we are dealing with. Managing such a mega-smart city poses many challenges at every level, from transport and communication through waste management and municipal management to housing. Each of these aspects requires wise management and the implementation of the best-developed solutions. It must not be forgotten that the city is also at risk of earthquakes as it is located at the junction of tectonic plates. This aspect means implementing solutions in the field of security, informing about the threat, and using a different way of construction. In addition, Istanbul is still a window to Asia and the largest transfer airport in this direction. The authorities of the city of Istanbul face new investments, implement new solutions, and search for new technologies every year. Without smart solutions, Istanbul would not be able to function as effectively as it does today.

The PPP is a long-term cooperation between the public sector and the private sector in the provision of services for society. As it is known, mega cities are characterized by a particularly high level of human capital; they are also distinguished by intensive scientific and research activity and the accumulation of innovative companies and public institutions. Among the features indicated as characteristics of the metropolis (enormous number of people, perfection of services, institutions and equipment, and the uniqueness and specificity of the place), there is also a multi-faceted innovative potential in the technical, economic, social, political, and cultural fields. Istanbul is undoubtedly a great metropolis. It functions in Turkey as a command post and continues to strive to become, among others, an international air cargo center or transfer center. A new port will be built in Istanbul to also become a launching point for cruises. Therefore, Istanbul has an adequate chance of becoming one of the regional—and later global—centers for entrepreneurship. This will make it an attractive arena for projects, entrepreneurs, and ecosystem actors from abroad.

In addition to the function of a business center and transport hub, Istanbul is also a center of science, knowledge, and innovation. Introducing PPP in cities such as Istanbul will

- Allow greater access to the expertise of private entities
- Improve access to innovative solutions (increasing the innovativeness of the economy)
- Encourage private entities to make financial investments that will enable the development of public employees
- Improve the quality of services through innovation and competition
- Encourage the economic development of the country/region
- Develop the tourism economy

In the further development of Istanbul as a smart city, it seems inevitable to use the experience of European PPP leaders (France, Germany, United Kingdom), as well as leaders of partnership development in the new EU Member States (Poland) (European Investment Bank, 2023) (see Figure 12.5).

In addition, it is important to initiate pilot projects first, limit PPP to large projects, and build substantive support for PPP in government administration units.

(Zegleń et al., 2017)

In recent years, the Turkish Council for Foreign Economic Relations (DEIK) has organized meetings with representatives from 42 different countries, and this effort is expected to expand in the coming years. The main objective of such activities is to make Istanbul a center of excellence for PPP (Istanbul PPP

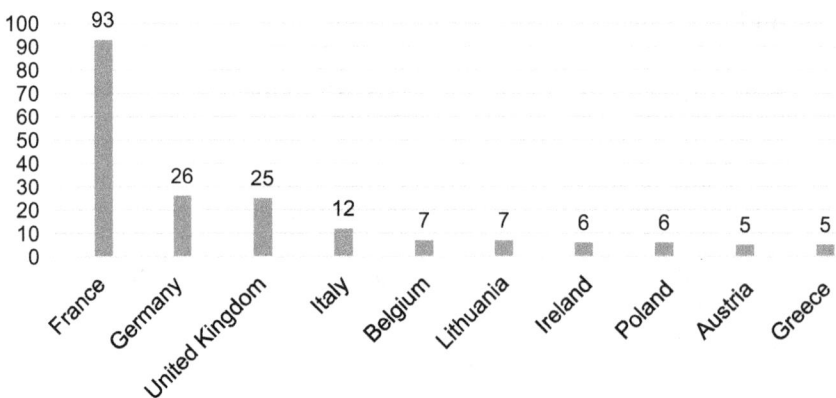

Figure 12.5 Evolution of the European public-private partnership market by country (2018–2022) in numbers.

Source: Author, based on: European PPP Expertise Centre — Market update 2022 (2023).

Week, 2023). Considering the funds spent on PPP objectives and the number of projects implemented, Istanbul has a chance to achieve such objectives.

The PPP opens up a number of opportunities for the city of Istanbul. A dynamically developing metropolis would probably not be able to develop organically as an intelligent city without financial support from private sources. Thanks to projects in PPP models, modern solutions are constantly being implemented, significantly affecting the quality of life in the city. At the same time, smart investments, such as a new airport, change the character of districts. Areas that have not been popular or inhabited so far are populated due to mega investments appearing there, such as airports or hospitals. Turkey as a country draws heavily from the opportunities offered by PPP, thanks to which a lot of projects have been implemented in a relatively short time, throwing Turkey as a country into a fast track of technological development.

References

Alhassani, A. and Almarri, K. (2020), "The impact of knowledge sharing and innovation performance on the organization–moderating by innovation awards", *Proceedings of Information Systems: 16th European, Mediterranean, and Middle Eastern Conference, EMCIS 2019*, Dubai, United Arab Emirates, December 9–10, pp. 512–523, Springer International Publishing.

Almarri, K. (2022), "The value for money factors and their interrelationships for smart city public–private partnerships projects", *Construction Innovation*. DOI: 10.1108/CI-01-2022-0020

Andon, P. (2012), "Accounting-related research in PPPs/PFIs: present contributions and future opportunities", *Accounting, Auditing & Accountability Journal*, Vol. 25, No. 5, pp. 876–924.

Broadbent, J., Laughlin, R. (2004), "PPPs: Nature, Development And Unanswered Questions", *Australian Accounting Review*, Vol. 14, No. 33, pp. 4–10.

Celikyay, H. (2017), "The Studies Through Smart Cities Model: The Case of Istanbul", *International Journal of Research in Business and Social Science*, Vol. 6, No. 1, pp. 149–163. DOI:10.20525/ijrbs.v6i1.713

Chih-Yao, H., Ren-Jye, D. (2019), "Evaluating Ancillary Business Scale for PPP-BOT Projects: A Social Housing BOT Case in Taiwan", *Sustainability*, Vol. 11, No. 5, pp. 1–17.

Delibasi, T. (2019), "Mega Istanbul Airport", *Network Industries Quarterly*, Vol. 21, No. 2, pp. 14–17.

Eren, F. (2018), "Top government hands-on megaproject management: the case of Istanbul's grand airport", *International Journal of Managing Projects in Business*, Vol. 12, No. 3, pp. 666–693. DOI: 10.1108/IJMPB-02-2018-0020

European Bank for Reconstruction and Development (2021), "Istanbul joins EBRD Green Cities urban sustainability programme", available at: https://www.ebrd.com/news/2021/istanbul-joins-ebrd-green-cities-urban-sustainability-programme.html (accessed 13th of May 2024).

European Court of Auditors (2018), "Public Private Partnerships in the EU: Widespread shortcomings and limited benefits", Special Report, available at: https://www.eca.europa.eu/Lists/ECADocuments/SR18_09/SR_PPP_EN.pdf, (accessed 13th of May 2024).

European Investment Bank (2023), "Market update Review of the European public-private partnership market in 2022", available at: https://www.eib.org/attachments/lucalli/20230009_epec_market_update_2022_en.pdf, (accessed 13th of May 2024).

Gobin-Rahimbux, B., Cadersaib, Z., Chooramun, N., Gooda Sahib-Kaudeer, N., Heenaye-Mamode Khan, M., Cheerkoot-Jalim, S., Kishnah, S. and Elaheeboccus, S. (2020), "A systematic literature review on ict architectures for smart mauritian local council", *Transforming Government: People, Process and Policy*, Vol. 14, No. 2, pp. 261–281. DOI: 10.1108/TG-07-2019-0062

Gonel, F. and Akinci, A. (2018), "How does ICT-use improve the environment? The case of Turkey", *World Journal of Science Technology and Sustainable Development*, Vol. 15, No. 1. DOI: 10.1108/WJSTSD-03-2017-0007

Istanbul Metropolitan Municipality (2023), "ISBAK AS", available at: https://ibb.istanbul/isbak-as/, (accessed 28th of December 2022).

Istanbul PPP Week, (2023), available at: https://istanbulpppweek.com/, (accessed 2nd of March 2023).

Jayasena, N., Chan, D.W.M. and Kumaraswamy, M. (2020), "A systematic literature review and analysis towards developing PPP models for delivering smart infrastructure", *Built Environment Project and Asset Management*. DOI: 10.1108/BEPAM-11-2019-0124

Jonek-Kowalska, I. and Kaźmierczak, J. (2021), Raport z ogólnopolskich badań nt. inteligentnych miast, Politechnika Śląska, available at: https://www.polsl.pl/rdnzj/wp-content/uploads/sites/803/2021/05/Raport-z-ogolnopolskich-badan-inteligentnych-miast.pdf, (accessed: 21st of December 2022).

Kuyucak, Sengur F. (2020), "Public-private partnerships in airports: The Turkish experience", *World Review of Intermodal Transportation Research*, Vol.9, No.3, pp. 217–244, DOI: 10.1504/WRITR.2020.10029515

Lam, P.T.I. and Yang, W. (2020), "Factors influencing the consideration of Public-Private Partnerships (PPP) for smart city projects: Evidence from Hong Kong", *Cities*, Vol. 99, p. 102606.

Liu T., Mostafa S., Mohamed S. and Nguyen T.S. (2020), "Emerging themes of public-private partnership application in developing smart city projects: A conceptual framework", *Built Environment Project and Asset Management*, Vol. 11, No. 1, pp. 138–156. DOI: 10.1108/BEPAM-12-2019-0142

Malek, M. and Gundaliya, P. (2021), "Value for money factors in Indian public-private partnership roadprojects: An exploratory approach", *Journal of Project Management*, Vol. 6, No. 1, pp. 23–32.

Malik, S. and Kaur, S. (2021), "Multi-dimensional public–private partnership readiness index: A sub-national analysis of India", *Transforming Government: People, Process and Policy*, Vol. 15, No. 4, pp. 483–511. DOI: 10.1108/TG-06-2020-0107

Ojasalo, J. and Kauppinen, H. (2016), "Collaborative innovation with external actors: an empirical study on open innovation platforms in smart cities", *Technology Innovation Management Review*, Vol. 6, pp. 49–60.

Pellegrino, R., Costantino, N. and Tauro, D. (2019), "Supply chain finance: A supply chain-oriented perspective to mitigate commodity risk and pricing volatility", *Journal of Purchasing and Supply Management*, Vol. 25, No. 2, pp. 118–133.

Pérez-delHoyo, R., Andújar-Montoya, M.D., Mora, H. and Gilart-Iglesias, V. (2019), "Unexpected consequences in the operation of urban environments", *Kybernetes*, Vol. 48, No. 2, pp. 253–264. DOI: 10.1108/K-02-2018-0096

Rathore, M.M., Paul, A., Ahmad, A. and Jeon, G. (2017), "IoT-based big data: from smart city towards next generation super city planning", *International Journal on Semantic Web and Information Systems*, Vol. 13, No. 1, pp. 28–47.

Report ISBAK, (2023), available at: https://isbak.istanbul/wp-content/uploads/2021/06/2020-ISBAK-Faaliyet-Raporu-RS-Min-Compressed.pdf, (accessed 23rd of November 2022).

Seema, A. and John, P. (2016), "Public–private partnerships for future urban infrastructure", *Proceedings of the Institution of Civil Engineers - Management, Procurement and Law*, Vol. 169, pp. 150–158.

Selim, A.M. and ElGohary, A.S. (2020), "Public–private partnerships (PPPs) in smart infrastructure projects: the role of stakeholders", *HBRC Journal*, Vol. 16, No. 1, pp. 317–333.

Selim, A.M., Yousef, P.H. and Hagag, M.R. (2018), "Smart infrastructure by (PPPs) within the concept of smart cities to achieve sustainable development", *International Journal of Critical Infrastructures*, Vol. 14, No. 2, pp. 182–198.

Singh, A. and Singla, A.R. (2021), "Constructing definition of smart cities from systems thinking view", *Kybernetes*, Vol. 50, No. 6, pp. 1919–1950.

Smart City Index 2023, (2022), available at: https://imd.cld.bz/IMD-Smart-City-Index-Report-2023; (accessed 27th of December 2022).

Tan, S.Y. and Taeihagh, A. (2020), "Smart city governance in developing countries: A systematic literature review", Switzerland, *Sustainability*, Vol. 12, No. 3, pp. 1–29.

Turkish Investment Office (2022), "Business services sector in Turkey", available at: www.invest.gov.tr/tr/library/Lists/InvestPublications/Business-Services-Industry.pdf#search=PPP (accessed 26th December 2022).

Visvizi, A. and Lytras, M.D. (eds) (2019), *Smart Cities: Issues and Challenges: Mapping Political, Social and Economic Risks and Threats*, Elsevier, ISBN: 9780128166390, https://www.elsevier.com/books/smart-cities-issues-and-challenges/lytras/978-0-12-816639-0et

Voorwinden, A. (2021), "The privatised city: Technology and publicprivate partnerships in the smart city", *Law, Innovation and Technology*, Vol. 13, No. 2, pp. 439–463. DOI: 10.1080/17579961.2021.1977213

Willems, T., Van Dooren, W. and Van Den Hurk, M. (2018), "PPP policy, depoliticisation, and anti-politics", *Partecipazione e Conflitto*, Vol. 10, No. 2, pp. 448–471.

Zegleń, P., Obodyński, M., Czarny, W., Zaborniak-Sobczak, M. and Matusikova, D., (2017), "Uwarunkowania rozwoju partnerstwa publiczno-prywatnego (PPP) jako możliwości wsparcia regionalnej gospodarki turystycznej", *Handel Wewnętrzny*, Vol. 2, No. 4(369), pp. 370–385, available at: https://cejsh.icm.edu.pl/cejsh/element/bwmeta1.element.desklight-482485b0-b22f-43ef-9c12-20bcdc9efd32/c/IBRKK-handel_wew_4-2017-t2.370-385.pdf (accessed 23rd of December 2022).

13 An alternative view on smart cities

Can small towns become smart?

Giovanni Baldi and Antonio Botti

13.1 Introduction

Pace is the key word to explain how much our way of life has changed over time, as history testifies. Due to advances in technology and evolution of the society and of the spatial patterns that surround us, everything becomes somewhat different every day (Cooper, 2005; Oztemel & Gursev, 2020). Since the turn of the century, the disruptive nature of technology has accelerated growth, changing, among other things, cities and how we live (Kagermann, 2015; Rucinski et al., 2017; Visvizi, 2022). However, around the world, rural areas or even small towns exist, be it the developed or developing countries, which – even if they exhibit a slow pace of living (in contrast to the high speed of the rest of the world) – still succeed in introducing and implementing novel and sustainable information and communication technology (ICT)-based solutions. This is precisely the crux of the issue that this chapter explores. In this kind of places, everyone leads their lives within the spatial frame of a city, large or small, where it is the community that serves as the point of reference for social and professional life (Anas et al., 1998; Angelidou, 2014; Visvizi & Lytras, 2018a, 2018b; Visvizi & Lytras, 2019a, 2019b). In this view, the city represents the boundaries of everyday life (Beauregard, 1990; Kipfer, 2018). However, to adapt to changes taking place in the society at large, brought by advances in ICT, as well as in the fabric of cities and also in small towns, spatial arrangements must also alter at this point (Polese et al., 2021; Visvizi & Troisi, 2022). Hence, it is important to study the processes of change of these slower small towns as they seek to become smarter and more sustainable. Does this mean that they become faster? Can small towns become smart?

To answer these questions, it is necessary to examine the main actors, i.e., the inhabitants, of a city. Indeed, for there to be smart and sustainable development, there needs to be an evolution, upskilling, and engagement of citizens (Huttunen et al., 2022). In a small city, this is even more difficult due to the barriers inherent within the small towns, mainly the level of education and age (Baldi & Megaro, 2023). Therefore, the question arises as to how the transition to a smart city is perceived by small town citizens. In order to explore the topic and answer the question, it was decided to interview the citizens of a small town

DOI: 10.1201/9781003415930-17

in southern Italy: Castellabate, in the province of Salerno. The argument in this chapter is structured as follows: after conducting a literature review in the next section, Section 13.3 describes the methodology, the context of the study, and the research results, in order to understand how the citizens of small towns are able to understand the slow process of smartening up their town and how it can also be redesigned spatially. In Section 13.4, discussions and conclusions are provided, together with the limitations and future implications of the research.

13.2 Literature review

13.2.1 *Fastness, slowness, and smartness*

As more people move from rural to urban areas in search of better opportunities, cities face numerous challenges associated with the rapid increase in population (Davis, 1965; Cohen, 2006; Agarwal et al., 2007; Andreasyan et al., 2021). With urbanization, cities often experience strain on essential resources like water and energy, leading to excessive consumption and potential scarcity. Additionally, the rise in population density contributes to higher levels of pollution, increased traffic congestion, and challenges in waste management. These issues can negatively impact the quality of life for urban residents, hindering access to vital services such as healthcare and education (Buhaug & Urdal, 2013; Haase, 2015; Visvizi et al., 2017; del Hoyo et al., 2021; Khan & Krishnan, 2021; Liebert & Wodarski, 2021). In a world where speed has been hailed as the catalyst for global and urban progress, there has traditionally been a negative association with slowness. This perception has particularly affected small, less-accessible urban centers in rural areas of developed or developing countries, often resulting in them being viewed as lagging behind their larger, developed, and smart counterparts. However, a paradigm shift is underway as scholars increasingly associate slowness with notions of well-being, child-friendliness, sustainability, and health. This shift in perspective has prompted numerous policymakers, planners, and politicians to revaluate and reimagine the concept of 'slow cities' (Ball, 2015; Tranter & Tolley, 2020). The concept of slow cities is centered around principles such as reducing vehicular traffic, promoting sustainable transportation modes like walking and cycling, establishing inviting and secure public spaces, and encouraging balanced and healthy lifestyles. By embodying these principles, slow cities offer a more humane environment, enabling individuals to enjoy a more relaxed pace of life, fostering social interaction, accessing high-quality services, and coexisting harmoniously with their surroundings (Mayer & Knox, 2006).

It is important to emphasize that slowness should not be misconstrued as stagnation or a lack of progress. Rather, it represents a more mindful and deliberate approach to urban planning and management. By acknowledging the value of slowness, cities can strive for sustainability and inclusivity and prioritize the well-being of their residents, particularly children and families (Raco et al., 2018; Knox, 2005). In conclusion, as the traditional emphasis on speed is being re-examined, scholars and professionals are increasingly recognizing the merits of embracing slowness as an alternative approach to conceiving and planning cities.

Slow cities offer advantages in terms of improved quality of life, sustainability, and health, paving the way for innovative policies and approaches in urban design (Xu et al., 2020; Raco et al., 2018). In this argument, the smart city is meant as a balanced approach that addresses both unsustainable speed and sustainable slowness in urban development. By leveraging advanced technologies, smart cities optimize efficiency and resource management, mitigating issues like congestion and excessive consumption. Simultaneously, they prioritize well-being, creating inclusive spaces and promoting a balanced pace of life. This dual focus ensures a higher quality of life and sustainable urban environments. Smart cities represent a paradigm shift toward holistic and mindful urban development.

13.2.2 *Smart cities*

To meet the challenges of digital and sustainable transition, both in the most developed and in the slowest areas, it might be useful to design or rethink cities from a smart perspective. The development of smart cities has gained prominence (Ersoy, 2017; Polese et al., 2018). A smart city harnesses the power of data, ICT, and innovative urban infrastructure to enhance the overall efficiency, sustainability, and liveability of urban areas. It leverages these technologies to improve governance, enhance quality of life, promote environmental sustainability, and drive economic growth (Sandeep et al., 2020; Ahmad et al., 2021; Benites & Simoes, 2021). The smart city model encompasses several dimensions that collectively contribute to the realization of its goals (Giffinger et al., 2007). These dimensions include

- Smart governance: This dimension focuses on leveraging technology to enhance the effectiveness and efficiency of urban governance. It involves utilizing digital platforms for citizen engagement, data-driven decision-making, and streamlined service delivery by local authorities.
- Smart living: This dimension revolves around improving the quality of life for residents by integrating technology into various aspects of urban living. It includes initiatives such as smart homes, intelligent transportation systems, and access to digital services that enhance convenience, safety, and overall well-being.
- Smart environment: This dimension emphasizes the sustainable management of urban resources and the preservation of the natural environment. It involves implementing measures to reduce pollution, optimize energy consumption, promote renewable energy sources, and enhance environmental monitoring and conservation efforts.
- Smart economy: This dimension focuses on fostering economic development and innovation within the city. It involves promoting entrepreneurship, supporting digital industries, encouraging research and development, and facilitating the integration of technology into traditional sectors to drive economic growth and create job opportunities.

By adopting smart city principles and strategies, urban areas can address the challenges brought about by urbanization and mitigate the slowness of a city. Smart technologies can optimize resource usage, improve infrastructure efficiency, and enhance the overall quality of life for residents. Furthermore, smart cities have the potential to contribute to sustainable development goals, mitigate the impacts of climate change, and create inclusive and resilient urban environments (Visvizi & Lytras, 2018b).

A smart city embodies an urban development model that leverages the collective human and technological capital within urban areas (Angelidou, 2014). It is conceptualized as an ultramodern urban environment that caters to the needs of businesses, institutions, and, most importantly, citizens (Khatoun & Zeadally, 2016). Extensive research on smart cities primarily focuses on the novel services enabled by ICTs, as well as their responsiveness and sustainability (Visvizi & Lytras, 2018b). The overarching goal of smart cities is to address citizen needs and improve the quality of life for residents (Shapiro, 2006). Scholarly contributions consistently highlight the significance of enhancing the quality of life in the context of smart cities. Liebert & Wodarski (2021) propose that smart cities represent urban development achieved through the optimized utilization of resources with the aid of ICTs and smart sustainability. Such an approach offers a multitude of benefits to residents, local authorities, and businesses, ultimately resulting in an improved quality of life within urban agglomerations. It is evident that a smart city encompasses a multidisciplinary concept, encompassing its information infrastructure and its capacity to manage resources and information effectively, all with the aim of enhancing the well-being of its inhabitants (Ramaprasad et al., 2017). Lazaroiu & Roscia (2012) also depict the smart city as a future-oriented challenge, a city model where technology serves people and contributes to the advancement of their economic and social quality of life. Similarly, Piro et al. (2014) describe an urban environment supported by pervasive information systems, delivering advanced and innovative services that further enhance the overall quality of life for citizens. In conclusion, the concept of a smart city revolves around the fundamental objective of improving the quality of life for its residents. By integrating advanced technologies, optimizing the use of resources, and ensuring sustainability, smart cities aim to provide a seamless urban experience that meets the needs of citizens and promotes their economic, social, and environmental well-being as if they were slow but smart cities.

13.2.3 *A multilevel structure for smart city urban redefinition*

The structure of a smart city focuses on representing the structure of the entire system, describing intra- and inter-system relationships, and defining guidelines and principles that govern design, development, and evolution over time. Generally, the architecture must meet certain standards, established by reasoning about the integration of technology and systems needed for the needs identified by stakeholders (local, regional, and national governments; citizen and public services; ICT for businesses, NGOs, etc.) and characteristics such as multilevel structure, interoperability, scalability, flexibility, fault-tolerance, manageability, resilience, standardization, security, privacy, and technology- or vendor-independent (Anthopoulos, 2017).

With reference to the multilevel structure, the International Telecommunication Union (ITU) divides the smart city into five main levels (Patrão et al., 2020):

- Natural environment: concerns the natural landscape of the city
- Hard infrastructure (non-ICT-based): buildings, infrastructure (roads, bridges, and telecommunications), and utilities (water, energy, and waste).
- Hard infrastructure (ICT-based): covers all the hardware, with which smart services are produced and delivered to end users (data centers, telecommunication networks, Internet of Things (IoT), and sensors).
- Smart services: include all services offered through the hard and soft infrastructure.
- Soft infrastructure: individuals and groups of people, business processes, software and data applications, with which smart services are delivered and implemented

(Lin et al., 2019)

Therefore, cities are entering a new digital and interconnected era through the use of new enabling technologies. In research and application, the drivers of the smart city transformation are IoT; big data and cloud computing; application programming interface (API) economy; cloud and fog computing; 5G connectivity; cybersecurity; artificial intelligence, machine learning, and deep learning; blockchain; and fintech (Yaqoob et al., 2017; Ahad et al., 2020; Anthony et al., 2020). These enabling innovations pervade all sectors of the economy, society, and thus the smart city in all micro-, meso-, and macro-level contexts (Visvizi & Lytras, 2018b). Technologies then fall on the so-called smart services of smart city which are smart water, smart energy, smart transportation, smart health, smart safety and emergency, smart environment and waste management, smart building and digital home, smart government, smart economy, and smart education and Tourism (Ismagilova et al., 2019; Walletzký et al., 2020).

13.2.4 Smart small and slow towns

Returning to the subject of slow cities, as we have said, these can be related to small towns. The definition of a small town can vary based on the context of geography, culture, and economy. Generally, small cities have populations ranging from a few thousand to tens of thousands. Some small towns are prosperous and lively, while others are simply a collection of smaller places, and some are at risk of becoming abandoned. Additionally, their spatial contexts vary widely, ranging from areas close to large urban areas to those far from other urban areas, which function as a regional center for a large rural hinterland, including small areas close to other small urban areas that form the backbone of a polycentric regional network (EU Commission, 2022). Mishkovsky et al. (2010) describe five general categories for small communities: gateway communities, resource-dependent communities, edge communities, traditional main street communities, and second home and retirement communities. Each type of community faces unique challenges related to limited access to services and resources, limited economic opportunities, and demographic changes such as an aging population and out-migration of young people. Despite these challenges, small cities can offer a unique quality of life characterized by a strong sense of community, a slower pace of life, and a connection to the natural environment that characterizes a slow city (D'Agostini & Fantini, 2008; Tranter & Tolley, 2020). Small towns may also have a rich cultural heritage, historic architecture, and natural attractions that contribute to their character and identity.

Recently, there has been an increase in interest in promoting sustainable development and resilience in small cities, with a focus on improving liveability, economic prosperity, and environmental sustainability. A liveable city is not only an economically competitive city offering numerous services and job opportunities, but also a healthy place that promotes the health, well-being, and equity of its citizens (Mohamed, 2023). Strategies to support local businesses and entrepreneurship, preserve cultural heritage and natural resources, and promote social inclusion and community engagement are critical in achieving this goal. There are studies in the literature analyzing the development of small towns, and successful cases mainly come from the United States. In general, to achieve sustainable development in small towns, it is first necessary to involve the citizenry also in the decision-making process (Matysiak & Peters, 2023). This can help ensure that the needs and priorities of the local population are considered in planning and development efforts. It is equally important to develop a strategic plan for urban development that considers the unique needs and characteristics of the local community. The idea is to capitalize on existing assets and create a win–win economy. This can help drive investment in infrastructure, economic development, and environmental sustainability (Cai, 2002; Arfini et al., 2019; Wang et al., 2019). Recently, the challenges for small cities also concern smart and green development with digital solutions that are adapted and built for specific cases such as applications for tourist flow management (Baldi et al., 2022), waste recycling systems (Massawe, 2014), and solutions in healthcare (Chen et al., 2021); water resource management (Telci & Aral, 2018; Farah & Shahrour, 2018), energy efficiency (Irwin et al., 2017), and public lighting (Furlan & Sipe, 2017; Cellucci et al., 2015); and emission reduction and energy saving (Senbel et al., 2013; Niemi et al., 2012).

Barriers to the development of smart projects that have jointly emerged in the literature relate to a problem with the adoption of technology by citizens due to their advanced age or low schooling (Baldi & Megaro, 2023) or the fact that they do not perceive the objective as relevant and not appropriate for the reality in which they live (Ševčík et al., 2022). In addition, there were often problems with resources, infrastructure, or citizens who are not properly involved in decision-making processes, especially young people who are less connected to the area than older people and are therefore more inclined to migrate to larger areas (Matysiak & Peters, 2023). It is necessary to rethink applications and services within environments in order to use them appropriately, as they affect the structural and infrastructural remodeling of city spaces, both physical and virtual (Polese et al., 2016; Vallicelli, 2018; Mundula et al., 2019; Aurigi & Odendaal, 2021; Afrin et al., 2021; Botti & Monda, 2022). As the literature reveals, there are issues with rural and small towns because of the aging population, lack of basic infrastructure, and desertification that already plague these places and which are addressed by the European Agenda 2030 (Rieniets, 2009; Camarero & Oliva, 2019; Mihai & Iatu, 2020; Tang et al., 2022).

These are all problems that discourage and potentially disable the installation and development of smart cities technologies that in order to solve them and have the same development, we need to investigate the perception and degree of acceptance and adoption of technologies by citizens in order to co-create solutions that are within their reach and truly citizen-centric (Oliveira & Campolargo, 2015; Sepasgozar et al., 2019; Habib et al., 2020; Dirsehan & van Zoonen, 2022; Baldi and Megaro, 2023). Most of these studies in this direction identify perceived usefulness and perceived ease of use as important predictors for the acceptance of smart city technologies. On the other hand, Troisi et al. (2022) looked at how users of technology in smart cities saw it, conceptualizing the multiple psychological and social beliefs, cultural customs, and rational reasons that go into the difficult adoption of technology and technological advancements. The majority of research, according to the study of the literature, is strictly quantitative, and the qualitative studies are typically cases of small cities that are examined holistically and conceptually, do not take the needs of the residents into account, and have never been conducted in Italy. Finally, no study of those mentioned refers to the needs and perceptions of small town citizens with respect to the smart development of their city and the spatial issues arising from the installation of new technologies within the urban and sometimes rural context (Campbell, 2013; Jara et al., 2015; Visvizi & Lytras, 2018b; Anindra et al., 2018; Nilssen, 2019; Ruohomaa et al., 2019; Dembski et al., 2020; Treude et al., 2022).

In the light of the literature review, this research question (**RQ**) was posed: *How do citizens of a small town perceive and conceive of the transition toward a smart city?*

Starting from this gap and answering this RQ, it was decided to interview citizens involved in an urban context of an Italian small, slow town to try to find out their needs, limitations, and trends for smart and sustainable town development with bottom-up approach starting right from the local community (Simon, 1991; Altieri & Masera, 1993; Fraser et al., 2006; Semeraro et al., 2020; Bours et al., 2022).

13.3   Methods and methodology

13.3.1   *Context of study and data collection*

The town taken as a reference is Castellabate, a larger than populated southern Italian municipality in the Cilento National Park, has about 8,000 inhabitants, and is known for its extremely touristic vocation based mainly on beach, landscape, and eno-gastronomic tourism, as well as for healthy and slow living. Through primary data collected at the town hall, most of the citizens are directly interested in tourist flows by owning or managing accommodation facilities or supporting them such as in the food and beverage, facility management, and personal services sectors (Baldi & Megaro, 2023).

Semi-structured interviews were conducted (Kamnuansilpa et al., 2020). The questions included a series of simple, straightforward questions about their perception and awareness of smart cities and how they might see their city changed and developed through enabling technologies, as well as questions to describe them from a socio-demographic perspective.

The questions were prepared digitally through MS-Forms and were administered by inviting in-person responses to citizens from September 2022 and February 2023 inside the city's municipal hall. Participants answered individually and independently through their smartphones without risk of being conditioned so that they already had the answers in text format and also to prevent them from answering in an altered or bogus way (Harrell & Bradley, 2009).

We stopped 124 people and obtained only 40 responses of which 5 were null; 35 responses were analyzed by content analysis through SAS® Viya® portfolio for key informant and word cloud to better visualize the trending topics, to answering the RQ, and directing future research (Bernard et al., 2016).

13.3.2   *Participants*

The interview participants on average are 50 years old as can be seen from tab.1. Just over half were women and 68% possess only a high school diploma. Moreover, the interviewees are all connected to tourism activities within the city: 78% own an accommodation facility to host tourists, 6% of them only work there, and finally 16% run a bar or restaurant. Finally, nearly 12% of those who run a lodging establishment do not know how to use the municipality's platform that is used to manage the flow of tourists, register guests, and collect and pay tourist taxes.

Almost all participants connected the conceptual and spatial development of their city toward a smart and sustainable city to the quality tourism improvement and growth, as they are all engaged in this sector.

Table 13.1 Participants' description

GENDER	Male 16 (45%)	Female 19 (55%)	Overall (N = 35)
Age	mean 50.4	means 50.7	median 50
EDUCATION			
Junior high school	3 (16%)	4 (18%)	7 (20%)
High school	12 (75%)	12 (64%)	24 (68%)
Bachelor's degree	0 (0%)	2 (13%)	2 (6%)
Master's degree	1 (9%)	1 (5%)	2 (6%)
OCCUPATION			
Facility owner	15 (94%)	12 (63%)	27 (78%)
Facility employee	0 (0%)	2 (10%)	2 (6%)
Food and beverage owner	1 (6%)	5 (27%)	6 (16%)
MUNICIPALITY TOURISM- PLATFORM USAGE			
Self-use	13 (81%)	18 (95%)	31 (88%)
With someone's help	3 (19%)	1 (5%)	4 (12%)

Source: Authors.

13.4   Findings

A few participants complained about the alleged irregularity and lack of guest registration of its contestants, such as Respondent 1 (R1):

> one can think of improving only after regularizing all illegal rentals so as to improve the 'average quality' of the tourists present. On August 16, only 4,700 tourists were registered in the face of no vacancies even in neighboring municipalities: a mockery of regular tour operators in a country that professes Quality Tourism (Table 13.1).

Either the R2: *"Let the scoundrels who rent off the books be identified, so tourism will be improved."*

Or the R3: *"Control must be carried out not on the certified facilities but on the off-the-books ones that house 10 people in squalid accommodations even within the gardens (in the alleys for example)."*

And finally, the R4: *"May control on wild and abusive rentals. Increased security by law enforcement."*

On the other hand, there are those who complain about the tourist tax (R5): *"Cancel the tourist tax since although it is a purpose tax, it is not used at all to improve tourism infrastructure."*

Or others who understand smart development of their city as improving the tourism offering in a digitized way, such as R6: *"A smart card for guests (parking, bus, Wi-Fi, UPDATED map) as in every TOURIST country"*

And R7: *"Real-time digital updating of available slots at free beaches and private lidos with increased spacing between slots."*

Many other participants associated the smart development of their city with improving network infrastructure and connectivity that is clearly lacking and weak or have more tourist information about events and attractions that we can share with guests, as R8 suggests: *"The municipality communicate on the platform, events, fairs, activities, services for tourists, closed roads, etc."*

Other respondents, however, suggested actions in which there is public–private cooperation in the provision of public works that would change the spatial configuration of the city toward a more sustainable urban environment for citizens and tourists.

Such as (R9):

Digitizing the entire municipal sector by offering maximum online services to people. Electric charging points for bikes as well. Platform for services accesses offered by third parties and which predominantly tourists may need. Agreements with sea-descent owners to allow controlled access via digital app for beach access especially for establishment-free stretches.

And R10: *"Municipally operated scooter and electric bike rentals (as in all cities now), free wi-fi, sustainable power for homes and facilities, modular benches...."*

Respondent 11: *"More car charging stations and more ATMs even in small hamlets."*

R12: *"More services to tourists and more parking areas."*

R13 as well:

Municipally operated and NOT PRIVATE scooter, scooter and electric bicycle rentals. Photovoltaic panels for clean, renewable energy generation aimed at efficient lighting and heating Smart urban transportation networks Safer public spaces for all ages ... our children are in the lurch, or in private hands

R14 also

Services for tourists, parking, cleanliness of beaches, clean and well-maintained free beaches. Maintenance of water drains. Create meeting points for young people and try to improve quality of tourism by offering adequate services. Improve the type of shopping, Too many little market stores: the Corso used to have only posh stores. There is a lack of local craft offerings.

And finally, R15: *"I would do something smarter on the issue of waste collection, always a big problem for both tourists and those who run accommodations."*

There are eventually some participants who would like more security and video surveillance, energy efficiency, and more effective digitization of the municipality's paperwork. A word cloud depicted in Figure 13.1 can summarize all the responses above and also the others that were not reported in the research results.

Figure 13.1 The responses word cloud.

13.5   Final remarks

Based on the results of the interviews and the analysis of the word cloud (Fig. 1), it becomes evident that the citizens of the small town examined in this chapter have a strong inclination toward, on the one hand, improving the economic quality of life for themselves and, on the other hand, enhancing the liveability of their guests, who contribute to the town's prosperity as tourists. This suggests the residents' commitment to the principles of well-being and a healthy environment in a city that, while appreciates slowness, also aspires to evolve in terms of efficiency and urban space planning.

Addressing the research question outlined in Section 13.3, the citizens interviewed identified several directions for the smart development of their city and the rethinking of its urban context. These directions align with a particular focus on environmental sustainability and align with the smart city architecture discussed in the literature review (Anthopoulos, 2017; Lin et al., 2019; Patrão et al., 2020). The identified directions include *connectivity, digitization, public infrastructures, smart mobility, energy efficiency, control* and *safety,* and a *multi-service platform.* These emerging factors correspond to the significance of hard infrastructure and smart services as primary drivers for the transformation of small towns into smart cities.

To ensure the successful implementation of smart city initiatives in small towns, collaborative public–private agreements that involve citizen engagement and a top-down governance approach are crucial (Bahaire & Elliott-White, 1999; Eshuis et al., 2014; Chan & Cao, 2015). The literature highlights the importance of tailoring digital technology solutions to address specific challenges that are unique to the characteristics of each area. This approach promotes resource, time, and energy efficiency, contributing to greater environmental and social sustainability. Notably, successful cases in the United States have emphasized the value of investing in local resources, including citizen talent, arts, and crafts and making use of underutilized municipal premises (UN-Habitat, 2023). Additionally, an intergenerational approach and open innovation processes have proven effective in engaging citizens and attracting external investment for implementing digital solutions that foster economic and sustainable development (Troisi et al., 2019; Hosseini et al., 2018; Clarinval et al., 2021; Kędra et al., 2023). These strategies can be applied to implement technology specifically tailored to small towns with a tourist focus.

This study contributes to the literature by addressing a gap that exists in research focusing primarily on citizen acceptance of smart city technologies in medium to large cities, while neglecting small and rural areas, which face distinct challenges outlined in the literature review. The findings provide valuable insights into various avenues for future research. For instance, it is important to explore other urban contexts using mixed qualitative and quantitative methods, examining the differences in the impact of smart city technologies on spaces in small and large cities. Furthermore, it is crucial to consider the influence of age on perceptions of continuity, slowness, digitization, and the adoption of smart city technologies. Understanding the differing perspectives among age groups can inform targeted strategies for engaging citizens in different demographic categories. Small towns possess unique cultural and tourist value, which coexists with the imperative of demarcation and the adoption of smart city approaches. Preserving the cultural and historical heritage of small towns is vital to maintain their distinctive identities and attract tourists. By carefully integrating smart solutions while respecting and enhancing the town's cultural assets, a balance can be achieved between continuity and change. This balance allows small towns to embrace both their cultural heritage and the benefits of smart city transformations.

While this study focused on a single small town with a strong tourism focus, which may have influenced the perspectives of the interview participants, it is important to acknowledge the limitations. Future research should encompass a broader range of small towns with different economic sectors and involve all stakeholders, including mayors, administrators, civil servants, local entrepreneurs, non-profit organizations, and citizens. This comprehensive approach will provide a more nuanced understanding of perceptions toward smart city transformations in small and slow cities and contribute to the development of a new theoretical framework. Longitudinal studies could also be conducted to observe changes in citizens' perceptions and needs within the dynamic ecosystem of smart cities.

References

Afrin, S., Chowdhury, F. J., & Rahman, M. M. (2021). COVID-19 pandemic: Rethinking strategies for resilient urban design, perceptions, and planning. *Frontiers in Sustainable Cities*, *3*, 668263.

Agarwal, S., Satyavada, A., Kaushik, S., & Kumar, R. (2007). Urbanization, urban poverty and health of the urban poor: status, challenges and the way forward. *Demography India*, 36(1), https://ssrn.com/abstract=3133050

Ahad, M. A., Paiva, S., Tripathi, G., & Feroz, N. (2020). Enabling technologies and sustainable smart cities. *Sustainable cities and society*, *61*, 102301.

Ahmad, A., Jeon, G., & Yu, C. W. (2021). Challenges and emerging technologies for sustainable smart cities. *Indoor and Built Environment*, *30*(5), 581–584.

Altieri, M. A., & Masera, O. (1993). Sustainable rural development in Latin America: Building from the bottom-up. *Ecological Economics*, *7*(2), 93–121.

Anas, A., Arnott, R., & Small, K. A. (1998). Urban spatial structure. *Journal of economic literature*, *36*(3), 1426–1464.

Andreasyan, N., Dorado, A. F. D., Colombo, M., Teran, L., Pincay, J., Nguyen, M. T., & Portmann, E. (2021, July). Framework for involving citizens in human smart city projects using collaborative events. In *2021 Eighth International Conference on eDemocracy & eGovernment (ICEDEG)* (pp. 103–109). IEEE.

Angelidou, M. (2014). Smart city policies: A spatial approach. *Cities*, *41*, S3–S11.

Anindra, F., Supangkat, S. H., & Kosala, R. R. (2018, October). Smart governance as smart city critical success factor (case in 15 cities in Indonesia). In *2018 International Conference on ICT for Smart Society (ICISS)* (pp. 1–6). IEEE.

Anthony, B., Petersen, S. A., & Helfert, M. (2020, November). Digital transformation of virtual enterprises for providing collaborative services in smart cities. In *Working conference on virtual enterprises* (pp. 249–260). Cham: Springer.

Anthopoulos, L. G. (2017). *Understanding smart cities: A tool for smart government or an industrial trick?* (Vol. 22, p. 293). Cham: Springer International Publishing.

Arfini, F., Antonioli, F., Cozzi, E., Donati, M., Guareschi, M., Mancini, M. C., & Veneziani, M. (2019). Sustainability, innovation and rural development: The case of Parmigiano-Reggiano PDO. *Sustainability*, 11(18), 4978.

Aurigi, A., & Odendaal, N. (2021). From "smart in the box" to "smart in the city": Rethinking the socially sustainable smart city in context. *Journal of Urban Technology*, *28*(1–2), 55–70.

Bahaire, T., & Elliott-White, M. (1999). Community participation in tourism planning and development in the historic city of York, England. *Current Issues in Tourism*, *2*(2–3), 243–276.

Baldi, G., & Megaro A. (2023). Smart small villages conceptualization based on the capabilities co-elevation for smart citizens. In *ITM Web of Conferences* (Vol. 51, p. 02004). EDP Sciences. https://doi.org/10.1051/itmconf/20235102004

Baldi, G., Megaro, A., Carrubbo, L. (2022). Small-town citizens' technology acceptance of smart and sustainable city development. *Sustainability*, 15(1), 325.

Ball, S. (2015). Slow cities. *Theme Cities: Solutions for Urban Problems*, 563–585.

Beauregard, R. A. (1990). Bringing the city back in American Planning Association. *Journal of the American Planning Association*, 56(2), 210.

Benites, A. J., & Simoes, A. F. (2021). Assessing the urban sustainable development strategy: An application of a smart city services sustainability taxonomy. *Ecological Indicators*, 127, 107734.

Bernard, H. R., Wutich, A., & Ryan, G. W. (2016). *Analyzing qualitative data: Systematic approaches*. SAGE publications.

Botti, A., & Monda, A. (2022). The evolution of the smart city in Italy: An empirical investigation on the importance of smart services. In: Visvizi, A., Troisi, O. (eds) *Managing Smart Cities*. Springer, Cham. https://doi.org/10.1007/978-3-030-93585-6_14

Bours, S. A., Wanzenböck, I., & Frenken, K. (2022). Small wins for grand challenges. A bottom-up governance approach to regional innovation policy. *European Planning Studies*, 30(11), 2245–2272.

Buhaug, H., & Urdal, H. (2013). An urbanization bomb? Population growth and social disorder in cities. *Global Environmental Change*, 23(1), 1–10.

Cai, L. A. (2002). Cooperative branding for rural destinations. *Annals of Tourism Research*, 29(3), 720–742.

Camarero, L., & Oliva, J. (2019). Thinking in rural gap: Mobility and social inequalities. *Palgrave Communications*, 5(1), 1–7.

Campbell, T. (2013). *Beyond smart cities: How cities network, learn and innovate.* Routledge.

Cellucci, L., Burattini, C., Drakou, D., Gugliermetti, F., Bisegna, F., de Lieto Vollaro, A., ... & Golasi, I. (2015). Urban lighting project for a small town: Comparing citizens and authority benefits. *Sustainability*, 7(10), 14230–14244

Chan, A. C. M., & Cao, T. (2015). Age-friendly neighbourhoods as civic participation: Implementation of an active ageing policy in Hong Kong. *Journal of Social Work Practice*, 29(1), 53–68.

Chen, D., Loboda, T. V., Silva, J. A., & Tonellato, M. R. (2021). Characterizing small-town development using very high resolution imagery within remote rural settings of Mozambique. *Remote Sensing*, 13(17), 3385.

Clarinval, A., Simonofski, A., Vanderose, B. and Dumas, B. (2021). Public displays and citizen participation: A systematic literature review and research agenda. *Transforming Government: People, Process and Policy*, 15(1), 1–35. https://doi.org/10.1108/TG-12-2019-0127

Cohen, B. (2006). Urbanization in developing countries: Current trends, future projections, and key challenges for sustainability. *Technology in Society*, 28(1–2), 63–80.

Cooper, T. (2005). Slower consumption reflections on product life spans and the "throwaway society". *Journal of Industrial Ecology*, 9(1–2), 51–67.

D'Agostini, L. R., & Fantini, A. C. (2008). Quality of life and quality of living conditions in rural areas: Distinctively perceived and quantitatively distinguished. *Social Indicators Research*, 89, 487–499

Davis, K. (1965). The urbanization of the human population. *Scientific American*, 213(3), 40–53.

del Hoyo, R.P., Visvizi, A. and Mora, H. (2021) 'Chapter 2 - Inclusiveness, safety, resilience, and sustainability in the smart city context', in A. Visvizi and R.P. del Hoyo (eds) *Smart Cities and the un SDGs*. Elsevier, pp. 15–28. Available at: https://doi.org/10.1016/B978-0-323-85151-0.00002-6

Dembski, F., Wössner, U., Letzgus, M., Ruddat, M., & Yamu, C. (2020). Urban digital twins for smart cities and citizens: The case study of Herrenberg, Germany. *Sustainability*, 12(6), 2307.

Dirsehan, T., & van Zoonen, L. (2022). Smart city technologies from the perspective of technology acceptance. *IET Smart Cities*, 4(3), 197–210.

Ersoy, A. (2017). Smart cities as a mechanism towards a broader understanding of infrastructure interdependencies. *Regional Studies, Regional Science*, 4(1), 26–31.

Eshuis, J., Klijn, E. H., & Braun, E. (2014). Place marketing and citizen participation: Branding as strategy to address the emotional dimension of policy making? *International Review of Administrative Sciences*, 80(1), 151–171.

EU Commission (2022). Small urban areas a foresight assessment to ensure a just transition. Available: https://cor.europa.eu/en/engage/studies/Documents/Small%20urban%20areas_a%20foresight%20assessment%20to%20ensure%20a%20just%20transition.pdf (Last access: 04/04/2023)

Farah, E., & Shahrour, I. (2018). Smart water technology for leakage detection: Feedback of large-scale experimentation. *Analog Integrated Circuits and Signal Processing*, 96, 235–242.

Fraser, E. D., Dougill, A. J., Mabee, W. E., Reed, M., & McAlpine, P. (2006). Bottom up and top down: Analysis of participatory processes for sustainability indicator identification as a pathway to community empowerment and sustainable environmental management. *Journal of Environmental Management*, 78(2), 114–127.

Furlan, R., & Sipe, N. (2017). Light rail transit (LRT) and transit villages in Qatar: A planning-strategy to revitalize the built environment of Doha. *Journal of Urban Regeneration and Renewal*, 10(4), 1–20.

Giffinger, R., Fertner, C., Kramar, H., & Meijers, E. (2007). City-ranking of European medium-sized cities. *Cent. Reg. Sci. Vienna UT*, 9(1), 1–12.

Haase, D. (2015). Reflections about blue ecosystem services in cities. *Sustainability of Water Quality and Ecology*, 5, 77–83.

Habib, A., Alsmadi, D., & Prybutok, V. R. (2020). Factors that determine residents' acceptance of smart city technologies. *Behaviour & Information Technology*, 39(6), 610–623.

Harrell, M. C., & Bradley, M. A. (2009). *Data collection methods. Semi-structured interviews and focus groups*. Rand National Defense Research Inst santa monica ca.

Hosseini, S., Frank, L., Fridgen, G., & Heger, S. (2018). Do not forget about smart towns: How to bring customized digital innovation to rural areas. *Business & Information Systems Engineering*, 60, 243–257.

Huttunen, S., Ojanen, M., Ott, A., & Saarikoski, H. (2022). What about citizens? A literature review of citizen engagement in sustainability transitions research. *Energy Research & Social Science*, 91, 102714.

Irwin, D., Iyengar, S., Lee, S., Mishra, A., Shenoy, P., & Xu, Y. (2017). Enabling distributed energy storage by incentivizing small load shifts. *ACM Transactions on Cyber-Physical Systems*, 1(2), 1–30

Ismagilova, E., Hughes, L., Dwivedi, Y. K., & Raman, K. R. (2019). Smart cities: Advances in research—An information systems perspective. *International Journal of Information Management*, 47, 88–100.

Jara, A. J., Sun, Y., Song, H., Bie, R., Genooud, D., & Bocchi, Y. (2015, March). Internet of Things for cultural heritage of smart cities and smart regions. In *2015 IEEE 29th International Conference on Advanced Information Networking and Applications Workshops* (pp. 668–675). IEEE.

Kagermann, H. (2015). Change through digitization—Value creation in the age of Industry 4.0. In: Albach, H., Meffert, H., Pinkwart, A., Reichwald, R. (eds) *Management of Permanent Change. Springer Gabler, Wiesbaden*. https://doi.org/10.1007/978-3-658-05014-6_2

Kamnuansilpa, P., Laochankham, S., Crumpton, C. D., & Draper, J. (2020). Citizen awareness of the smart city: A study of Khon Kaen, Thailand. *The Journal of Asian Finance, Economics and Business*, 7(7), 497–508.

Kędra, A., Maleszyk, P., & Visvizi, A. (2023). Engaging citizens in land use policy in the smart city context. *Land Use Policy*, 129, 106649.

Khan, A., & Krishnan, S. (2021). Citizen engagement in co-creation of e-government services: A process theory view from a meta-synthesis approach. *Internet Research*, 31(4), 1318–1375.

Khatoun, R., & Zeadally, S. (2016). Smart cities: concepts, architectures, research opportunities. *Communications of the ACM*, 59(8), 46–57.

Kipfer, S. (2018). Pushing the limits of urban research: Urbanization, pipelines and counter-colonial politics. *Environment and Planning D: Society and Space*, 36(3), 474–493.

Knox, P. L. (2005). Creating ordinary places: Slow cities in a fast world. *Journal of Urban Design*, 10(1), 1–11.

Lazaroiu, G. C., & Roscia, M. (2012). Definition methodology for the smart cities model. *Energy*, 47(1), 326–332.

Liebert, F., & Wodarski, K. (2021). Conceptual framework for measuring awareness and needs of city residents towards a smart city. *Organizacja i Zarządzanie: Kwartalnik naukowy*.

Lin, C. C., Liu, W. Y., & Lu, Y. W. (2019). Three-dimensional internet-of-things deployment with optimal management service benefits for smart tourism services in forest recreation parks. *IEEE Access*, 7, 182366–182380.

Massawe, E., Legleu, T., Vasut, L., Brandon, K., & Shelden, G. (2014). Voluntary approaches to solid waste management in small towns: A case study of community involvement in household hazardous waste recycling. *Journal of Environmental Health*, 76(10), 26–33.

Matysiak, I., & Peters, D. J. (2023). Conditions facilitating aging in place in rural communities: The case of smart senior towns in Iowa. *Journal of Rural Studies*, 97, 507–516.

Mayer, H., & Knox, P. L. (2006). Slow cities: Sustainable places in a fast world. *Journal of Urban Affairs*, 28(4), 321–334.

Mihai, F. C., & Iatu, C. (2020). Sustainable rural development under Agenda 2030. *Sustainability Assessment at the 21st century*, 9–18.

Mishkovsky, N., Dalbey, M., Bertaina, S., Read, A., & McGalliard, T. (2010). *Putting smart growth to work in rural communities*. International City/County Management Association (ICMA).

Mohamed, A.S.Y. (2023). Livable city: Broadening the smart city paradigm, insights from Saudi Arabia. In: Visvizi, A., Troisi, O., Grimaldi, M. (eds) *Research and innovation forum 2022. RIIFORUM 2022. Springer proceedings in complexity*. Springer. https://doi.org/10.1007/978-3-031-19560-0_15

Mundula, L., Balletto, G., & Borruso, G. (2019, July). The 'Dark Side' of the smartness. In *International conference on computational science and its applications* (pp. 253–268). Springer.

Niemi, R., Mikkola, J., & Lund, P. D. (2012). Urban energy systems with smart multi-carrier energy networks and renewable energy generation. *Renewable Energy*, 48, 524–536.

Nilssen, M. (2019). To the smart city and beyond? Developing a typology of smart urban innovation. *Technological Forecasting and Social Change*, 142, 98–104.

Oliveira, Á., & Campolargo, M. (2015, January). From smart cities to human smart cities. In *2015 48th Hawaii international conference on system sciences* (pp. 2336–2344). IEEE.

Oztemel, E., & Gursev, S. (2020). Literature review of Industry 4.0 and related technologies. *Journal of Intelligent Manufacturing*, 31(1), 127–182.

Patrão, C., Moura, P., & Almeida, A. T. D. (2020). Review of smart city assessment tools. *Smart Cities*, 3(4), 1117–1132.

Piro, G., Cianci, I., Grieco, L. A., Boggia, G., & Camarda, P. (2014). Information centric services in smart cities. *Journal of Systems and Software*, 88, 169–188.

Polese, F., Botti, A., Monda, A., & Grimaldi, M. (2018). Smart city as a service system: A framework to improve smart service management. *Journal of Service Science and Management*, 12(01), 1.

Polese, F., Tommasetti, A., Vesci, M., Carrubbo, L., & Troisi, O. (2016, May). Decision-making in smart service systems: A viable systems approach contribution to service science advances. In *International conference on exploring services science* (pp. 3–14). Springer.

Polese, F., Troisi, O., Grimaldi, M., & Loia, F. (2021). Reinterpreting governance in smart cities: An ecosystem-based view. In A. Visvizi and R.P. del Hoyo (eds) *Smart Cities and the un SDGs*. Elsevier, pp. 71–89.

Raco, M., Durrant, D., & Livingstone, N. (2018). Slow cities, urban politics and the temporalities of planning: Lessons from London. *Environment and Planning C: Politics and Space*, 36(7), 1176–1194.

Ramaprasad, A., Sánchez-Ortiz, A., & Syn, T. (2017, September). A unified definition of a smart city. In *International conference on electronic Government* (pp. 13–24). Springer.

Rieniets, T. (2009). Shrinking cities: Causes and effects of urban population losses in the twentieth century. *Nature and Culture*, 4(3), 231–254.

Rucinski, A., Garbos, R., Jeffords, J., & Chowdbury, S. (2017, September). Disruptive innovation in the era of global cyber-society: With focus on smart city efforts. In *2017 9th IEEE international conference on intelligent data acquisition and advanced computing Systems: Technology and applications (IDAACS)* (Vol. 2, pp. 1102–1104). IEEE.

Ruohomaa, H., Salminen, V., & Kunttu, I. (2019). *Towards a smart city concept in small cities.* https://osuva.uwasa.fi/handle/10024/10704

Sandeep, V., Honagond, P. V., Pujari, P. S., Kim, S. C., & Salkuti, S. R. (2020). A comprehensive study on smart cities: Recent developments, challenges and opportunities. *Indonesian Journal of Electrical Engineering and Computer Science*, 20(2), 575–582.

Semeraro, T., Zaccarelli, N., Lara, A., Sergi Cucinelli, F., & Aretano, R. (2020). A bottom-up and top-down participatory approach to planning and designing local urban development: Evidence from an urban university center. *Land*, 9(4), 98.

Senbel, M., van der Laan, M., Kellett, R., Girling, C., & Stuart, J. (2013). Can form-based code help reduce municipal greenhouse gas emissions in small towns? The case of Revelstoke, British Columbia. *Canadian Journal of Urban Research*, 22(1), 72–92.

Sepasgozar, S. M., Hawken, S., Sargolzaei, S., & Foroozanfa, M. (2019). Implementing citizen centric technology in developing smart cities: A model for predicting the acceptance of urban technologies. *Technological Forecasting and Social Change*, 142, 105–116.

Ševčík, M., Chaloupková, M., Zourková, I., & Janošíková, L. (2022). Barriers to the Implementation of Smart Projects in Rural Areas, Small Towns, and the City in Brno Metropolitan Area. *European Countryside*, 14(4), 675–695.

Shapiro, J. M. (2006). Smart cities: Quality of life, productivity, and the growth effects of human capital. *The review of economics and statistics*, 88(2), 324–335.

Simon, H. A. (1991). The architecture of complexity. In *Facets of systems science* (pp. 457–476). Springer.

Tang, J., Xiong, K., Chen, Y., Wang, Q., Ying, B., & Zhou, J. (2022). A review of village ecosystem vulnerability and resilience: Implications for the rocky desertification control. *International Journal of Environmental Research and Public Health*, 19(11), 6664.

Telci, I. T., & Aral, M. M. (2018). Optimal energy recovery from water distribution systems using smart operation scheduling. *Water*, 10(10), 1464.

Tranter, P., & Tolley, R. (2020). *Slow cities: Conquering our speed addiction for health and sustainability.* Elsevier.

Treude, M., Schüle, R., & Haake, H. (2022). Smart sustainable cities—Case study Südwestfalen Germany. *Sustainability*, 14(10), 5957.

Troisi, O., Fenza, G., Grimaldi, M., & Loia, F. (2022). Covid-19 sentiments in smart cities: The role of technology anxiety before and during the pandemic. *Computers in Human Behavior*, 126, 106986.

Troisi, O., Grimaldi, M., & Monda, A. (2019). Managing smart service ecosystems through technology: How ICTs enable value cocreation. *Tourism Analysis*, 24(3), 377–393.

UN-HABITAT (2023). Small town development approaches. Available at: https://unhabitat.org/small-town-development-approaches-the-global-urban-economic-dialogue-series-series-title (Last access: 04/04/2023).

Vallicelli, M. (2018). Smart cities and digital workplace culture in the global European context: Amsterdam, London and Paris. *City, Culture and Society*, 12, 25–34.

Visvizi, A. (2022). Computers and human behavior in the smart city: Issues, topics, and new research directions. *Computers in Human Behavior*, 140, 107596.

Visvizi, A., & Lytras, M. (Eds.). (2019a). Smart cities: Issues and challenges: Mapping political, social and economic risks and threats.

Visvizi, A., & Lytras, M. D. (2018a). It's not a fad: Smart cities and smart villages research in European and global contexts. *Sustainability*, 10(8), 2727.

Visvizi, A., & Lytras, M. D. (2018b). Rescaling and refocusing smart cities research: From mega cities to smart villages. *Journal of Science and Technology Policy Management*, 9(2), 134–145.

Visvizi, A., & Lytras, M. D. (2019b). Sustainable smart cities and smart villages research: Rethinking security, safety, well-being, and happiness. *Sustainability*, 12(1), 215.

Visvizi, A., Mazzucelli, C., & Lytras, M. (2017). Irregular migratory flows: Towards an ICTs' enabled integrated framework for resilient urban systems. *Journal of Science and Technology Policy Management*, 8(2), 227–242.

Visvizi, A., & Troisi, O. (2022). Effective management of the smart city: An outline of a conversation. In: Visvizi, A., Troisi, O. (eds) *Managing Smart Cities*. Springer, Cham. https://doi.org/10.1007/978-3-030-93585-6_1

Walletzký, L., Carrubbo, L., Toli, A. M., Ge, M., & Romanovská, F. (2020, July). Multi-contextual view to smart city architecture. In *International conference on applied human factors and ergonomics* (pp. 306–312). Springer.

Wang, X., Liu, S., Sykes, O., & Wang, C. (2019). Characteristic development model: A transformation for the sustainable development of small towns in China. *Sustainability*, 11(13), 3753.

Xu, G., Zhou, Z., Jiao, L., & Zhao, R. (2020). Compact urban form and expansion pattern slow down the decline in urban densities: A global perspective. *Land Use Policy*, 94, 104563.

Yaqoob, I., Hashem, I. A. T., Mehmood, Y., Gani, A., Mokhtar, S., & Guizani, S. (2017). Enabling communication technologies for smart cities. *IEEE Communications Magazine*, 55(1), 112–120.

Index

Pages in *italics* refer to figures and pages in **bold** refer to tables.

For Product Safety Concerns and Information please contact our EU
representative GPSR@taylorandfrancis.com
Taylor & Francis Verlag GmbH, Kaufingerstraße 24, 80331 München, Germany

www.ingramcontent.com/pod-product-compliance
Lightning Source LLC
Chambersburg PA
CBHW060255220326
41598CB00027B/4113